Collins

M000167316

2023 GUIDE
to the
NIGHT SKY

SOUTHERN HEMISPHERE

Storm Dunlop and Wil Tirion

Published by Collins
An imprint of HarperCollins*Publishers*
Westerhill Road, Bishopbriggs, Glasgow G64 2QT
www.harpercollins.co.uk

HarperCollins*Publishers*
1st Floor, Watermarque Building, Ringsend Road, Dublin 4, Ireland

© HarperCollins Publishers 2022
Text and illustrations © Storm Dunlop and Wil Tirion
Photographs © see acknowledgements page 110.

Collins ® is a registered trademark of HarperCollins Publishers Ltd

All rights reserved. No part of this publication may be reproduced, stored in a retrieval system,
or transmitted, in any form or by any means, electronic, mechanical, photocopying, recording
or otherwise without the prior written permission of the publisher and copyright owners.

The contents of this publication are believed correct at the time of printing.
Nevertheless the publisher can accept no responsibility for errors or omissions,
changes in the detail given or for any expense or loss thereby caused.

HarperCollins does not warrant that any website mentioned in this title will be provided uninterrupted,
that any website will be error free, that defects will be corrected, or that the website or the server that
makes it available are free of viruses or bugs. For full terms and conditions please refer to the site
terms provided on the website.

A catalogue record for this book is available from the British Library

ISBN 978-0-00-853257-4

10 9 8 7 6 5 4 3 2 1

Printed in China

If you would like to comment on any aspect of this book, please contact us at the above address or online.
e-mail: collinsmaps@harpercollins.co.uk

facebook.com/CollinsAstronomy

@CollinsAstro

MIX
Paper from
responsible sources
FSC
www.fsc.org
FSC™ C007454

This book is produced from independently certified
FSC™ paper to ensure responsible forest management.

For more information visit: www.harpercollins.co.uk/green

Contents

Introduction

The aim of this Guide is to help people find their way around the night sky, by showing how the stars that are visible change from month to month and by including details of various events that occur throughout the year. The objects and events described may be observed with the naked eye, or nothing more complicated than a pair of binoculars.

The conditions for observing naturally vary over the course of the year. During the summer, twilight may persist throughout the night and make it difficult to see the faintest stars. There are three recognized stages of twilight: civil twilight, when the Sun is less than 6° below the horizon; nautical twilight, when the Sun is between 6° and 12° below the horizon; and astronomical twilight, when the Sun is between 12° and 18° below the horizon. Full darkness occurs only when the Sun is more than 18° below the horizon. During nautical twilight, only the very brightest (navigation) stars are visible. During astronomical twilight, the faintest stars visible to the naked eye may be seen directly overhead, but are lost at lower altitudes. At Sydney, full darkness persists for about six hours at mid-summer. Even at Christchurch, NZ (not shown), full darkness

lasts about four hours. By contrast, as far south as Cape Horn, at mid-summer nautical twilight persists, so only the very brightest stars are visible.

Another factor that affects the visibility of objects is the amount of moonlight in the sky. At Full Moon, it may be very difficult to see some of the fainter stars and objects, and even when the Moon is at a smaller phase it may seriously interfere with visibility if it is near the stars or planets in which you are interested. A full lunar calendar is given for each month and may be used to see when nights are likely to be darkest and best for observation.

The celestial sphere

All the objects in the sky (including the Sun, Moon and stars) appear to lie at some indeterminate distance on a large sphere, centred on the Earth. This *celestial sphere* has various reference points and features that are related to those of the Earth. If the Earth's rotational axis is extended, for example, it points to the North and South Celestial Poles, which are thus in line with the North and South Poles on Earth. Similarly, the *celestial equator* lies in the same plane as the Earth's equator,

The duration of twilight throughout the year at Sydney and Cape Horn.

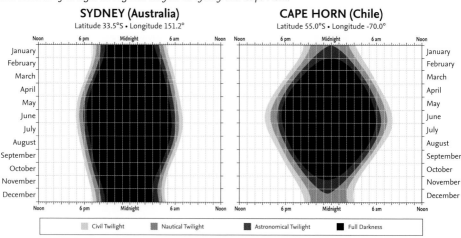

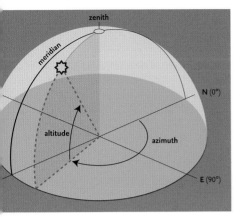

Measuring altitude and azimuth on the celestial sphere.

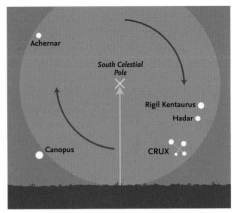

The altitude of the South Celestial Pole equals the observer's latitude.

and divides the sky into northern and southern hemispheres. Because this Guide is written for use in the southern hemisphere, the area of the sky that it describes includes the whole of the southern celestial hemisphere and those portions of the northern that become visible at different times of the year. Stars in the far north, however, remain invisible throughout the year, and are not included.

It is useful to know some of the special terms for various parts of the sky. As seen by an observer, half of the celestial sphere is invisible, below the horizon. The point directly overhead is known as the **zenith**, and the (invisible) one below one's feet as the **nadir**. The line running from the north point on the horizon, up through the zenith and then down to the south point is the **meridian**. This is an important invisible line in the sky, because objects are highest in the sky, and thus easiest to see, when they cross the meridian. Objects are said to **transit**, when they cross this line in the sky.

In this book, reference is frequently made in the text and in the diagrams to the standard compass points around the horizon. The position of any object in the sky may be described by its **altitude** (measured in degrees above the horizon), and its **azimuth** (measured in degrees from north 0°, through east 90°,

south 180° and west 270°). Experienced amateurs and professional astronomers also use another system of specifying locations on the celestial sphere, but that need not concern us here, where the simpler method will suffice.

The celestial sphere appears to rotate about an invisible axis, running between the North and South Celestial Poles. The location (i.e., the altitude) of the Celestial Poles depends entirely on the observer's position on Earth or, more specifically, their latitude. The charts in this book are produced for the latitude of 35°S, so the South Celestial Pole (SCP) is 35° above the southern horizon. The fact that the SCP is fixed relative to the horizon means that all the stars within 35° of the pole are always above the horizon and may, therefore, always be seen at night, regardless of the time of year. The southern circumpolar region is an ideal place to begin learning the sky, and ways to identify the circumpolar stars and constellations will be described shortly.

The ecliptic and the zodiac

Another important line on the celestial sphere is the Sun's apparent path against the background stars – in reality the result of the Earth's orbit around the Sun. This is known as the **ecliptic**. The point where the Sun, apparently moving along the ecliptic,

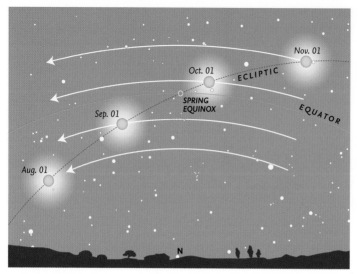

The Sun crossing the celestial equator at the September equinox (spring equinox in the southern hemisphere.).

crosses the celestial equator from south to north is known as the (southern) autumn equinox, which occurs on March 20 or 21. At this time and at the (southern) spring equinox, on September 22 or 23, when the Sun crosses the celestial equator from north to south, day and night are almost exactly equal in length. (There is a slight difference, but that need not concern us here.) The March equinox is currently located in the constellation of Pisces, and is important in astronomy because it defines the zero point for a system of celestial coordinates, which is, however, not used in this Guide.

The Moon and planets are to be found in a band of sky that extends 8° on either side of the ecliptic. This is because the orbits of the Moon and planets are inclined at various angles to the ecliptic (i.e., to the plane of the Earth's orbit). This band of sky is known as the zodiac and, when originally devised, consisted of twelve **constellations**, all of which were considered to be exactly 30° wide. When the constellation boundaries were formally established by the International Astronomical Union in 1930, the exact extent of most constellations was altered and, nowadays, the ecliptic passes through thirteen constellations.

Because of the boundary changes, the Moon and planets may actually pass through several other constellations that are adjacent to the original twelve.

The constellations

Since ancient times, the celestial sphere has been divided into various constellations, most dating back to antiquity and usually associated with certain myths or legendary people and animals. Nowadays, the boundaries of the constellations have been fixed by international agreement and their names (in Latin) are largely derived from Greek or Roman originals. Some of the names of the most prominent stars are of Greek or Roman origin, but many are derived from Arabic names. Many bright stars have no individual names and, for many years, stars were identified by terms such as 'the star in Hercules' right foot'. A more sensible scheme was introduced by the German astronomer Johannes Bayer in the early seventeenth century. Following his scheme – which is still used today – most of the brightest stars are identified by a Greek letter followed by the genitive form of the constellation's Latin name. An example is the Pole Star, also known as Polaris and α

Ursae Minoris (abbreviated α UMi). The Greek alphabet is shown on page 110 and a list of all the constellations that may be seen from latitude 35°S, together with abbreviations, their genitive forms and English names is on page 109. Other naming schemes exist for fainter stars, but are not used in this book.

Asterisms

Apart from the constellations (88 of which cover the whole sky), certain groups of stars, which may form a part of a larger constellation or cross several constellations, are readily recognizable and have been given individual names. These groups are known as *asterisms*, and the most famous (and well-known to northern observers) is the 'Plough', the common name for the seven brightest stars in the constellation of Ursa Major, the Great Bear. The names and details of some asterisms mentioned in this book are given in the list on page 110.

Magnitudes

The brightness of a star, planet or other body is frequently given in magnitudes (mag.). This is a mathematically defined scale where larger numbers indicate a fainter object. The scale extends beyond the zero point to negative numbers for very bright objects. (Sirius, the brightest star in the sky is mag. -1.4.) Most observers are able to see stars to about mag. 6, under very clear skies.

The Moon

Although the daily rotation of the Earth carries the sky from east to west, the Moon gradually moves eastwards by approximately its diameter (about half a degree) in an hour. Normally, in its orbit around the Earth, the Moon passes above or below the direct line between Earth and Sun (at New Moon) or outside the area obscured by the Earth's shadow (at Full Moon). Occasionally, however, the three bodies are more-or-less perfectly aligned to give an *eclipse*: a solar eclipse at New Moon or a lunar eclipse at Full Moon. Depending on the exact circumstances, a solar eclipse may be merely partial (when the Moon does not cover the whole of the Sun's disk); annular (when the Moon is too far from Earth in its orbit to appear large enough to hide the whole of the Sun); or total. Total and annular eclipses are visible from very restricted areas of the Earth, but partial eclipses are normally visible over a wider area.

Somewhat similarly, at a lunar eclipse, the Moon may pass through the outer zone of the Earth's shadow, the *penumbra* (in a penumbral eclipse, which is not generally perceptible to the naked eye), so that just part of the Moon is within the darkest part of the Earth's shadow, the *umbra* (in a partial eclipse); or completely within the umbra (in a total eclipse). Unlike solar eclipses, lunar eclipses are visible from large areas of the Earth.

Occasionally, as it moves across the sky, the Moon passes between the Earth and individual planets or distant stars, giving rise to an *occultation*. As with solar eclipses, such occultations are visible from restricted areas of the world.

The planets

Because the planets are always moving against the background stars, they are treated in some detail in the monthly pages and information is given when they are close to other planets, the Moon or any of five bright stars that lie near the ecliptic. Such events are known as *appulses* or, more frequently, as *conjunctions*. (There are technical differences in the way these terms are defined – and should be used – in astronomy, but these need not concern us here.) The positions of the planets are shown for every month on a special chart of the ecliptic.

The term conjunction is also used when a planet is either directly behind or in front of the Sun, as seen from Earth. (Under normal circumstances it will then be invisible.) The conditions of most favourable visibility depend on whether the planet is one of the two known as *inferior planets* (Mercury and Venus) or one of the three *superior planets* (Mars, Jupiter and Saturn) that are covered in detail. (Some details of the fainter superior planets, Uranus and

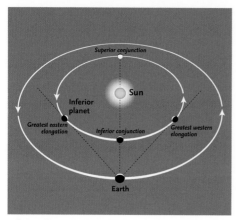

Inferior planet.

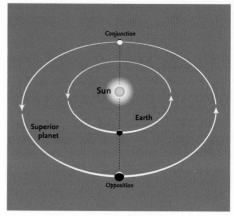

Superior planet.

Neptune, are included in this Guide, and special charts for both are given on page 25.)

The inferior planets are most readily seen at eastern or western **elongation**, when their angular distance from the Sun is greatest. For superior planets, they are best seen at **opposition**, when they are directly opposite the Sun in the sky, and cross the meridian at local midnight.

It is often useful to be able to estimate angles on the sky, and approximate values may be obtained by holding one hand at arm's length. The various angles are shown in the diagram, together with the separations of the various stars in and around Orion.

Meteors

At some time or other, nearly everyone has seen a **meteor** – a 'shooting star' – as it flashed across the sky. The particles that cause meteors – known technically as 'meteoroids' – range in size from that of a grain of sand (or even smaller) to the size of a pea. On any night of the year there are occasional meteors, known as **sporadics**, that may travel in any direction. These occur at a rate that is normally between three and eight in an hour. Far more important, however, are **meteor showers**, which occur at fixed periods of the year, when the Earth encounters a trail of particles left behind by a comet or, very occasionally, by a minor planet (asteroid). Meteors always appear to diverge from a single point on the sky, known as the **radiant**, and the radiants of major showers are shown on the charts. Meteors that come from a circular area 8° in diameter around the radiant are classed as belonging to the particular shower. All others that do not come from that area are sporadics (or, occasionally from another shower that is active at the same time). A list of the major meteor showers is given on page 31.

Although the positions of the various shower radiants are shown on the charts, looking directly at the radiant is not the most effective way of seeing meteors. They are most likely to be noticed if one is looking about 40–45° away from the radiant position. (This is approximately two hand-spans as shown in the diagram for measuring angles.)

Other objects

Certain other objects may be seen with the naked eye under good conditions. Some were given names in antiquity – Praesepe is one example – but many are known by what are called 'Messier numbers', the numbers in a catalogue of nebulous objects compiled by Charles Messier in the late eighteenth century. Some, such as the Andromeda Galaxy, M31, and the Orion Nebula, M42, may be seen by the naked eye, but all those given in the list will benefit from the use of binoculars.

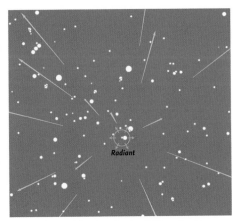

Meteor shower (showing the Geminids radiant).

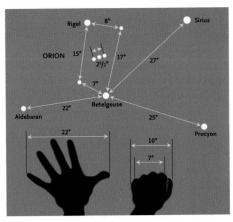

Measuring angles in the sky.

Apart from galaxies, such as M31, which contain thousands of millions of stars, there are also two types of cluster: open clusters, such as M45, the Pleiades, which may consist of a few dozen to some hundreds of stars; and globular clusters, such as Omega Centauri, which are spherical concentrations of many thousands of stars. One or two gaseous nebulae (emission nebulae), consisting of gas illuminated by stars within them, are also visible. The Orion Nebula, M42, is one, and is illuminated by the group of four stars, known as the Trapezium, which may be seen within it by using a good pair of binoculars.

Some interesting objects.

Messier / NGC	Name	Type	Constellation	Maps (months)
—	47 Tucanae	globular cluster	Tucana	All year
—	Hyades	open cluster	Taurus	Oct. – Feb.
—	Melotte 111 (Coma Cluster)	open cluster	Coma Berenices	Mar. – Jun.
M3	—	globular cluster	Canes Venatici	Mar. – Jul.
M4	—	globular cluster	Scorpius	Mar. – Sep.
M8	Lagoon Nebula	gaseous nebula	Sagittarius	Apr. – Oct.
M11	Wild Duck Cluster	open cluster	Scutum	May – Oct.
M13	Hercules Cluster	globular cluster	Hercules	May – Aug.
M15	—	globular cluster	Pegasus	Jul. – Nov.
M20	Trifid Nebula	gaseous nebula	Sagittarius	Apr. – Oct.
M22	—	globular cluster	Sagittarius	Apr. – Oct.
M27	Dumbbell Nebula	planetary nebula	Vulpecula	Jun. – Oct.
M31	Andromeda Galaxy	galaxy	Andromeda	Sep. – Dec.
M35	—	open cluster	Gemini	Nov. – Mar.
M42	Orion Nebula	gaseous nebula	Orion	Oct. – Apr.
M44	Praesepe	open cluster	Cancer	Dec. – Apr.
M45	Pleiades	open cluster	Taurus	Oct. – Feb.
M57	Ring Nebula	planetary nebula	Lyra	Jun. – Sep.
M67	—	open cluster	Cancer	Dec. – May
IC 2602	Southern Pleiades	open cluster	Carina	All year
NGC 2070	Tarantula Nebula	emission nebula	Dorado (LMC)	All year
NGC 3242	Ghost of Jupiter	planetary nebula	Hydra	Jan. – Jun.
NGC 3372	Eta Carinae Nebula	gaseous nebula	Carina	All year
NGC 4755	Jewel Box	open cluster	Crux	All year
NGC 5139	Omega Centauri	globular cluster	Centaurus	Jan. – Sep.

The Southern Circumpolar Constellations

The southern circumpolar constellations are the key to to starting to identify the constellations. For anyone in the southern hemisphere, they are visible at any time of the year, and nearly everyone is familiar with the striking pattern of four stars that make up the constelllation of **Crux** (the Southern Cross), and also the two nearby bright stars **Rigil Kentaurus** and **Hadar** (α and β Centauri respectively). This pattern of stars is visible throughout the year for most observers, although for observers farther north, the stars may become difficult to see, low on the horizon in the southern spring, espcially in the months of October and November.

Crux

The distinctive shape of the constellation of **Crux** is usually easy to identify, although some people (especially northerners unused to the southern sky) may wrongly identify the slightly larger False Cross, formed by the stars **Aspidiske** and **Avior** (ι and ε Carinae respectively) plus **Markeb** and **Alsephina** (κ and δ Velorum). The dark patch of the Coalsack (a dark cloud of obscuring dust) is readily visible on the eastern side of Crux between **Acrux** and **Mimosa** (α and β Crucis).

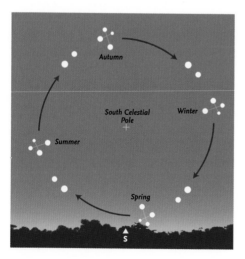

A line through **Gacrux** (γ Crucis) and **Acrux** (α Crucis) points approximately in the direction of the South Celestial Pole, crossing the faint constellations of **Musca** and **Chamaeleon**. Although there is no bright star close to the South Celestial Pole (and even the constellation in which it lies, **Octans**, is faint) an idea of its location helps to identify the region of sky that is always visible. (The altitude of the South Celestial Pole is equal to the observer's latitude south of the equator.) The basic triangular shape of Octans itself is best found by extending a line from **Peacock** (α Pavonis) through β Pavonis, by about the same distance as that between the stars.

Centaurus

Although Crux is a distinctive shape, **Centaurus** is a large, rather straggling constellation, with one notable object, the giant, bright globular cluster **Omega Centauri** (the brightest globular in the sky), which lies towards the north, on the line from **Hadar** (β Centauri) through ε Centauri. Rather than locating the South Celestial Pole from Crux, a better indication is the line, at right angles to the line between Hadar and Rigil Kentaurus that passes across the brightest star in **Circinus** (α).

Carina

Apart from the two stars that form part of the False Cross, the constellation of Carina is, like Centaurus, a large, sprawling constellation. It contains one striking open cluster, the **Southern Pleiades**, and a remarkable emission nebula, the **Eta Carinae Nebula**. The second brightest star in the sky (after Sirius) is **Canopus**, α Carinae, which lies far away to the west.

The Magellanic Clouds

On the opposite side of the South Celestial Pole to Crux and Centaurus lie the two Magellanic Clouds. The **Small Magellanic Cloud** (SMC) lies to one side of the relatively inconspicuous, triangular constellation of **Hydrus**, but is actually within the constellation of **Tucana**.

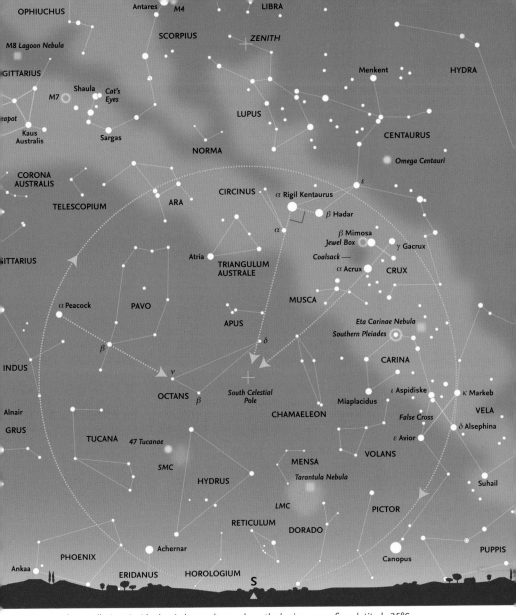

The stars and constellations inside the circle are always above the horizon, seen from latitude 35°S.

Nearby is another bright globular cluster, **47 Tucanae.** Hydrus itself is also easily identified from the star, **Achernar,** α Eridani, the rather isolated brilliant star at the southern end of **Eridanus,** which wanders a long way south, having begun at the foot of Orion.

The **Large Magellanic Cloud** (LMC) lies within the faint constellation of **Dorado.** Not only is it a large satellite galaxy of the Milky Way Galaxy, but it contains the large, readily visible, **Tarantula Nebula,** an emission nebula that is a major star-forming region.

The Summer Constellations

The summer sky is dominated by the constellation of **Orion**, with its three 'Belt' stars, as well as by the three bright stars: **Betelgeuse** (α Orionis), **Sirius** (α Canis Majoris), and **Procyon** (α Canis Minoris), which together form the prominent (Southern) 'Summer Triangle'. To the west of Orion is **Taurus**, with orange **Aldebaran** (α Tauri). A line from **Bellatrix** (γ Orionis) through Aldebaran points to the striking **Pleiades** cluster (M45). (The line of three 'Belt' stars indicates the same general path and, in the other direction, points towards Sirius.) A line from **Mintaka** (the westernmost star of the 'Belt') through brilliant **Rigel** (β Orionis) points in the direction of **Achernar** (α Eridani). One from **Alnitak** (at the other end of the 'Belt') through **Saiph** (κ Orionis) indicates the direction of **Canopus** (α Carinae), the second brightest star in the whole sky. A line from Betelgeuse through Sirius points in the general direction of the 'False Cross' in **Vela** and **Carina**.

In the opposite (northern) side of Orion, a line from **Meissa** (λ Orionis) through **Elnath** (β Tauri) leads to **Capella** (α Aurigae), and one from **Betelgeuse** through **Alhena** (γ Geminorum) points to **Pollux** (β Geminorium), the southernmost of the distinctive Castor/Pollux pair. A line across Orion from Bellatrix to Betelgeuse, if extended greatly, indicates the general direction of **Regulus** (α Leonis).

Six of the bright stars that have been mentioned, in different constellations: Capella, Aldebaran, Rigel, Sirius, Procyon and Pollux form what, for northern observers, is sometimes known as the 'Winter Hexagon'. Pollux is accompanied to the northwest by the slightly fainter star of **Castor** (α Geminorum), the second 'Twin'.

Several of the stars in this region of the sky show distinctive tints: Betelgeuse (α Orionis) is reddish, Aldebaran (α Tauri) is orange and Rigel (α Orionis) is blue-white.

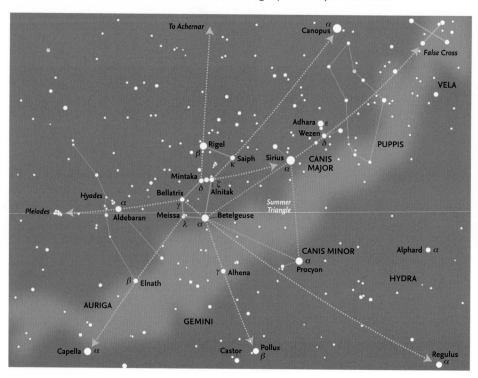

The Autumn Constellations

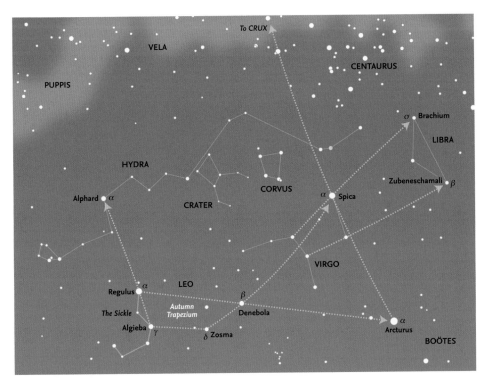

During the autumn season, the principal constellation is the zodiacal constellation of *Leo*, whose principal star **Regulus** (α Leonis) is one of the five bright stars that may occasionally be occulted by the Moon. Regulus forms the 'dot' of the upside-down 'question mark' of stars, known as the 'Sickle'. The body of Leo itself is sometimes known as the 'Trapezium' (here, the 'Autumn Trapezium'), and its sides may be used to find other objects in the sky.

A line extended from **Algiebra** (γ Leonis), the second star in the 'Sickle', through Regulus points to **Alphard** (α Hydrae) the brightest star in **Hydra**, the largest of the 88 constellations, which sprawls a long way across the sky from its distinctively shaped 'head' of stars, southwest of Regulus. A line from Regulus through **Denebola** (β Leonis) at the other end of the constellation of Leo points towards **Arcturus** (α Boötis), the brightest star in the

northern hemisphere of the sky. Another line, down the 'back' of Leo, from **Zosma** (δ Leonis) through Denebola points in the general direction of **Spica** (α Virginis) the principal star in the constellation of **Virgo**.

Although Spica is the brightest star in the zodiacal constellation of Virgo, the remainder of the constellation is not remarkable, consisting of a rough quadrilateral of stars with fainter lines extending outwards. The two sides of the main quadrilateral, if extended, point eastwards towards the two principal stars, **Zubeneschamali** and **Brachium** (β and σ Librae, respectively), of the small zodiacal constellation of **Libra**. The two small constellations of **Corvus** and **Crater** lie between Virgo and the long stretch of Hydra. A line from Arcturus through Spica, if extended far to the south across the sky, points to the distinctive constellation of **Crux**.

The Winter Constellations

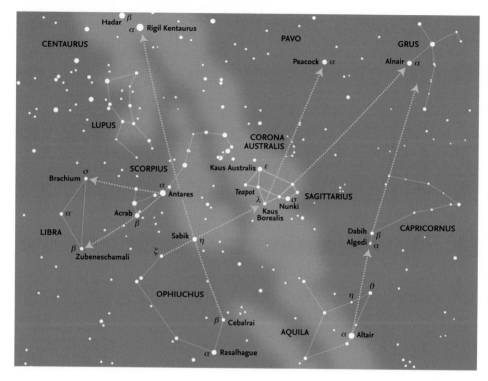

During the winter season, the Milky Way runs right across the sky, and the distinctive constellation of **Scorpius** is clearly visible, with bright red **Antares** (α Scorpii) and the 'sting', trailing behind the star to the west. Scorpius used to be a much larger constellation until the small zodiacal constellation of **Libra** was formed from some of its stars. The two sides of the 'fan' of stars immediately east of Antares, if extended, point to the two principal stars of Libra, **Zubeneschamali** and **Brachium** (β and σ Librae, respectively) on the southern and northern sides of the constellation.

North of Scorpius lies the large constellation of **Ophiuchus**. The ecliptic runs through the southern portion of Ophiuchus, and the Sun spends far more time in that constellation than it does in neighbouring Scorpius. A line from **Cebalrai** (β Ophiuchi) in the north of the constellation, through **Sabik** (η Ophiuchi), if extended right across

Scorpius, carries you to the two bright stars of **Centaurus**, **Rigil Kentaurus** and **Hadar** (α and β Centauri, respectively). Between Libra and Centaurus lies the straggling constellation of **Lupus** with no distinct pattern of stars.

A line from the fainter star ζ Ophiuchi through **Sabik** points towards **Kaus Borealis** (λ Sagittarii) the 'lid' of the 'Teapot' of **Sagittarius**. From there two lines radiate (on the west) across **Corona Australis** towards **Peacock** (α Pavonis) and (on the east) **Alnair** (α Gruis) in the small constellation of **Grus**.

Also north of **Scorpius** is the constellation of **Aquila**, with its brightest star, **Altair** (α Aquilae). The diamond shape of the 'wings' of Aquila may be used to locate **Algedi** (α Capricorni), a prominent double star in **Capricornus**. A line through **Dabih** (β Capricorni) and the western side of that basically triangular constellation, when extended, also points in the direction of the constellation of **Grus**.

The Spring Constellations

In spring, the constellation of **Pegasus** and the Great Square of Pegasus are prominent in the north. One star in the Square, **Alpheratz**, at the northwestern corner, actually belongs to the neighbouring constellation of **Andromeda**. A diagonal line across the Square from Alpheratz (α Andromedae) through **Markab** (α Pegasi) points to **Sadalmelik** (α Aquarii), just to the west of the 'Y-shaped' asterism of the 'Water Jar' in the zodiacal constellation of **Aquarius**. Further extended, that line carries you to the centre of the triangular constallation of **Capricornus**.

The constellation of **Pisces** consists of two lines of stars running to the south and east of the Great Square. In mythological representations, the two fish are linked by ribbons, tied together at **Alrescha** (α Piscium). The distinctive asterism of **The Circlet** lies at the western end of the southern line of stars. The two north-south lines of the Great Square may be used as guides to other

constellations. The western line from **Scheat** (β Pegasi) through Markab, if extended about three times, crosses Aquarius and points to the bright, fairly isolated star, **Fomalhaut** (α Piscis Austrini) in the constellation of **Piscis Austrinus** and then onward to **Tiaki** (β Gruis) in the centre of a cross of stars within **Grus**. The eastern line, from Alpheratz through **Algenib** (γ Pegasi), extended about three-and-a-half times, indicates **Ankaa** (α Phoenicis) in constellation of **Phoenix**.

The line between Markab and Algenib may be extended as an arc to lead to **Menkar** (α Ceti) in the 'tail' of Cetus. A line from Menkar through **Diphda** (β Ceti) also points to Fomalhaut in Piscis Austrinus. Diphda and Fomalhaut form the base of an almost perfect, large isoceles triangle, with **Achernar** (α Eridani) at the other apex. **Eridanus** itself is a long, trailing constellation that begins far to the north, near **Rigel** in **Orion**.

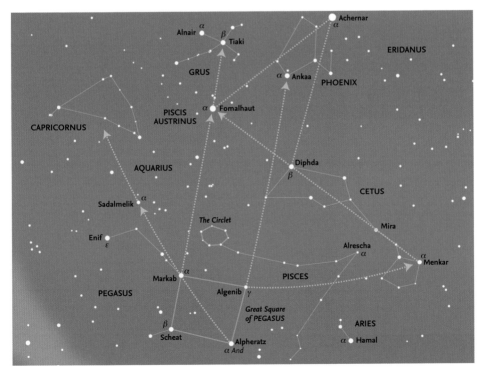

The Moon

The monthly pages include diagrams showing the phase of the Moon for every day of the month, and also indicate the day in the *lunation* (or *age* of the Moon), which begins at New Moon. Although the main features of the surface – the light highlands and the dark maria (seas) – may be seen with the naked eye, far more features may be detected with the use of binoculars or any telescope. The many craters are best seen when they are close to the *terminator* (the boundary between the illuminated and the non-illuminated areas of the surface), when the Sun rises or sets over any particular region of the Moon and the crater walls or central peaks cast strong shadows. Most features become difficult to see at Full Moon, although this is the best time to see the bright ray systems surrounding certain craters. Accompanying the Moon map on the following pages is a list of prominent features, including the days in the lunation when features are normally close to the terminator and thus easiest to see. A few bright features

such as Linné and Proclus, visible when well illuminated, are also listed. One feature, Rupes Recta (the Straight Wall) is readily visible only when it casts a shadow with light from the east, appearing as a light line when illuminated from the opposite direction.

The dates of visibility vary slightly through the effects of *libration*. Because the Moon's orbit is inclined to the Earth's equator and also because it moves in an ellipse, the Moon appears to rock slightly from side to side (and nod up and down). Features near the *limb* (the edge of the Moon) may vary considerably in their location and visibility. (This is easily noticeable with Mare Crisium and the craters Tycho and Plato.) Another effect is that at crescent phases before and after New Moon, the normally non-illuminated portion of the Moon receives a certain amount of light, reflected from the Earth. This *Earthshine* may enable certain bright features (such as Aristarchus, Kepler and Copernicus) to be detected even though they are not illuminated by sunlight.

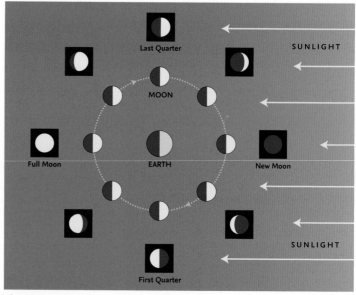

The Moon phases. *During its orbit around the Earth we see different portions of the illuminated side of the Moon's surface.*

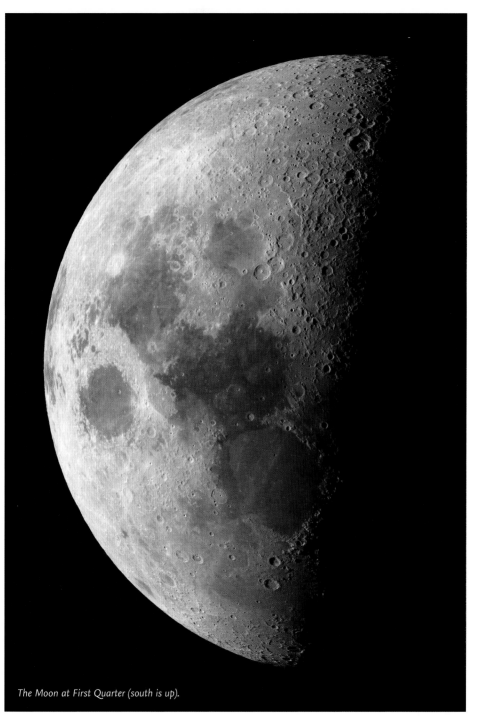

The Moon at First Quarter (south is up).

Map of the Moon

Abulfeda	6:20
Agrippa	7:21
Albategnius	7:21
Aliacensis	7:21
Alphonsus	8:22
Anaxagoras	9:23
Anaximenes	11:25
Archimedes	8:22
Aristarchus	11:25
Aristillus	7:21
Aristoteles	6:20
Arzachel	8:22
Atlas	4:18
Autolycus	7:21
Barrow	7:21
Billy	12:26
Birt	8:22
Blancanus	9:23
Bullialdus	9:23
Bürg	5:19
Campanus	10:24
Cassini	7:21
Catharina	6:20
Clavius	9:23
Cleomedes	3:17
Copernicus	9:23
Cyrillus	6:20
Delambre	6:20
Deslandres	8:22
Endymion	3:17
Eratosthenes	8:22
Eudoxus	6:20
Fra Mauro	9:23
Fracastorius	5:19
Franklin	4:18
Gassendi	11:25
Geminus	3:17
Goclenius	4:18
Grimaldi	13-14:27-28
Gutenberg	5:19
Hercules	5:19
Herodotus	11:25
Hipparchus	7:21
Hommel	5:19
Humboldt	3:15
Janssen	4:18
Julius Caesar	6:20
Kepler	10:24
Landsberg	10:24
Langrenus	3:17
Letronne	11:25
Linné	6
Longomontanus	9:23

The numbers indicate the age of the Moon when features are usually best visible.

SOUTH

Moretus, Manzinus, Curtius, Jacobi, Hommel, Vlacq, Baco, Cuvier, Pitiscus, Licetus, Maginus, Barocius, Saussure, Fabricius, Maurolycus, Nasireddin, Rheita, Metius, Riccius, Buch, Stöfler, Orontius, Furnerius, Büsching, Rabbi Levi, Nonius, Walther, Stevinus, Zagut, Aliacensis, Reichenbach, Pontanus, Werner, Regiomontanus, Humboldt, Snellius, Piccolomini, Apianus, Purbach, Petavius, Sacrobosco, Blanchinus, Azophi, Playfair, La Caille, Thebit, Santbech, Abenezra, Fracastorius, Beaumont, Geber, Arzachel, Vendelinus, Colombo, MARE NECTARIS, Catharina, Tacitus, Almanon, Airy, Argelander, Cyrillus, Abulfeda, Alphonsus, Kapteyn, Mädler, Theophilus, Albategnius, Goclenius, Ptolemaeus, Langrenus, Gutenberg, Isidorus, Hipparchus, Herschel, MARE FECUNDITATIS, Capella, Torricelli, Hypatia, Flammarion, Messier · A, Delambre, Rhaeticus, MARE SPUMANS, Sabine, Ritter, Godin, SINUS MEDII, Apollonius, Taruntius, Agrippa, Triesnecker, Bode, MARE UNDARUM, Firmicus, Arago, Hyginus, MARE TRANQUILLITATIS, Condorcet, Picard, PALUS SOMNI, Plinius, Menelaus, Manilius, MARE VAPORUM, MARE CRISIUM, Vetruvius, Macrobius, Bessel, Cocon, Le Monnier, Linné, PALUS PUTREDINIS, Cleomedes, MARE SERENITATIS, Autolycus, Archimedes, Hahn, Burckhardt, Chacornac, Posidonius, Montes Apenninus, Berosus, Gauss, Geminus, Montes Caucasus, Aristillus, Franklin, LACUS SOMNIORUM, Cassini, Cepheus, Atlas, Hercules, Bürg, LACUS MORTIS, Eudoxus, Montes Alpes, Mercurius, Aristoteles, Alpes Vallis, Endymion, MARE FRIGORIS, W. Bond, Arnold, Barrow, Meton, Goldschmidt

MARE AUSTRALE, Vallis Rheita, Montes Haemus

NORTH

18

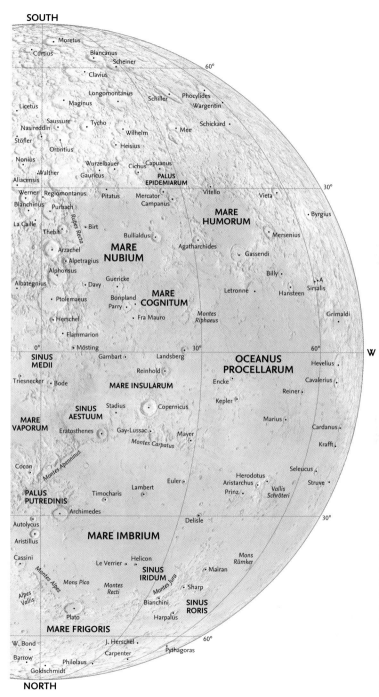

SOUTH

NORTH

Macrobius	4:18
Mädler	5:19
Maginus	8:22
Manilius	7:21
Mare Crisium	2-3:16-17
Maurolycus	6:20
Mercator	10:24
Metius	4:18
Meton	6:20
Mons Pico	8:22
Mons Piton	8:22
Mons Rümker	12:26
Montes Alpes	6-8:21
Montes Apenninus	8
Orontius	8:22
Pallas	8:22
Petavius	3:17
Philolaus	9:23
Piccolomini	5:19
Pitatus	8:22
Pitiscus	5:19
Plato	8:22
Plinius	6:20
Posidonius	5:19
Proclus	14:18
Ptolemaeus	8:22
Purbach	8:22
Pythagoras	12:26
Rabbi Levi	6:20
Reinhold	9:23
Rima Ariadaeus	6:20
Rupes Recta	8
Saussure	8:22
Scheiner	10:24
Schickard	12:26
Sinus Iridum	10:24
Snellius	3:17
Stöfler	7:21
Taruntius	4:18
Thebit	8:22
Theophilus	5:19
Timocharis	8:22
Triesnecker	6-7:21
Tycho	8:22
Vallis Alpes	7:21
Vallis Schröteri	11:25
Vlacq	5:19
Walther	7:21
Wargentin	12:27
Werner	7:21
Wilhelm	9:23
Zagut	6:20

Eclipses in 2023

Lunar eclipses

There are two lunar eclipses in 2023. Neither is particularly striking. The first, on May 5, is a penumbral eclipse, so will not be readily seen by the naked eye. This eclipse will occur over the Pacific, eastern Asia and Africa. The second, on October 28, is a very small partial eclipse. Only east Africa and parts of the Indian Ocean will see just a very tiny portion of the Moon enter the Earth's umbra.

Solar eclipses

There are also two solar eclipses in 2023. The eclipse of April 20 is known as a hybrid eclipse. The path runs from the Indian Ocean, across Indonesia, and out into the Pacific. A total eclipse (mid-eclipse 04:16) occurs southeast of the island of Flores. Because of the curvature of the Earth, the distance to the Moon is slightly greater over most of the path, where the eclipse will be annular. The second eclipse, on October 14 is annular, and the path runs across the United States, beginning in the northwest, then across Central America (greatest eclipse at 17:59 UT is off the coast of Nicaragua) and then northern South America (Colombia and northern Brazil).

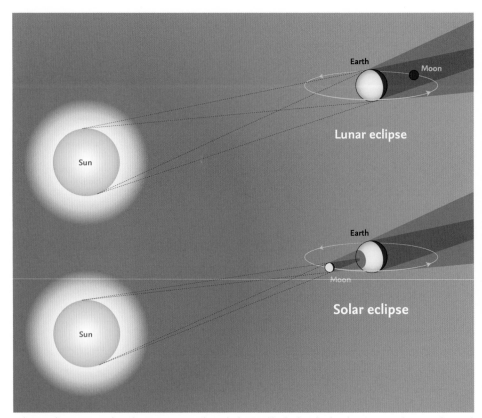

A lunar eclipse occurs when the Moon passes through the Earth's shadow (top). When it passes between the Earth and the Sun (bottom) there is a solar eclipse.

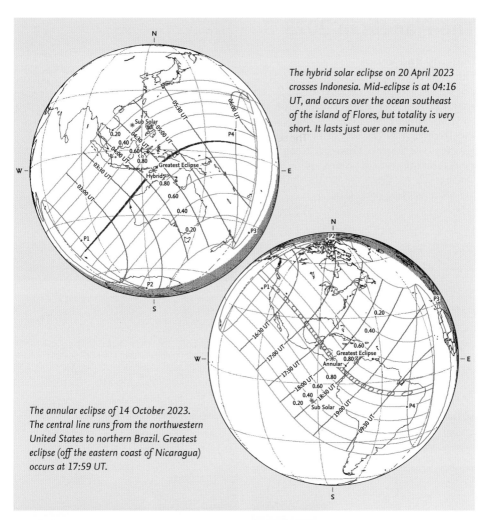

The hybrid solar eclipse on 20 April 2023 crosses Indonesia. Mid-eclipse is at 04:16 UT, and occurs over the ocean southeast of the island of Flores, but totality is very short. It lasts just over one minute.

The annular eclipse of 14 October 2023. The central line runs from the northwestern United States to northern Brazil. Greatest eclipse (off the eastern coast of Nicaragua) occurs at 17:59 UT.

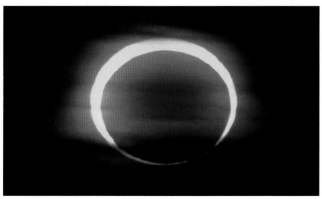

An annular eclipse of the Sun, photographed from California, towards the end of the eclipse at sunset over the Pacific.

The Planets in 2023

Mercury and Venus

Although **Mercury** comes to greatest elongation on six occasions in 2023, on only four is it reasonably visible above the horizon: January 30, May 29, August 10 and December 4. On one of these (August 10), Venus is nearby. **Venus** itself comes to elongation on two occasions: June 4 and October 23. All these events are shown on the accompanying diagrams.

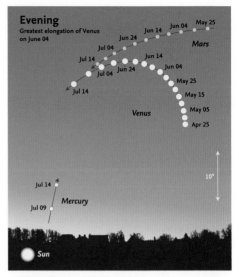

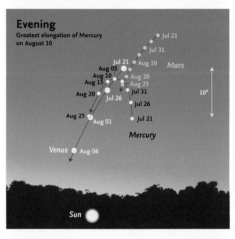

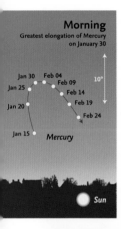

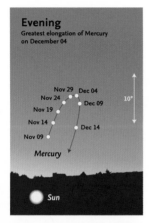

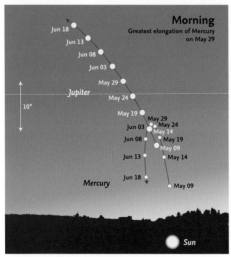

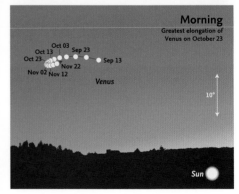

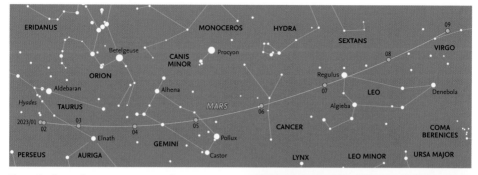

The path of Mars from January to September 2023. Later in the year it is too close to the Sun to be observed.

Mars

Because its orbit lies outside that of the Earth, so taking longer to complete an orbit, *Mars* does not come to opposition every year. Instead it often remains in the sky for many months at a time, slowly moving along the ecliptic. This is the case in 2023, when there is no opposition, but the planet is visible from January to September, as shown on the large chart here. It begins the year in **Taurus** at mag. -1.2. It passes right across the sky, slowly fading to mag. 1.8 in July and August, when it is in **Leo**, and ending the year at mag. 1.4, when it is too close to the Sun to be visible.

Oppositions occur during a period of retrograde motion, when the planet appears to move westwards against the pattern of distant stars. There is no opposition in 2024. Mars actually begins to retrograde on 14 December 2024, and comes to its next opposition on 16 January 2025.

Because of its eccentric orbit, which carries it at very differing distances from the Sun (and Earth), not all oppositions of Mars are equally favourable for observation. The relative positions of Mars and the Earth are shown here. It will be seen that the opposition of 2018 was very close and thus favourable for observation, and that of 2020 was also reasonably good. By comparison, opposition in 2027 will be at a far greater distance, so the planet will appear much smaller.

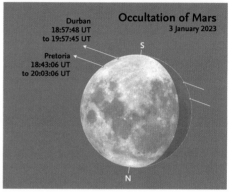

There are two occultations of Mars in January 2023. The occultation on 3 January will be visible from southern Africa. The paths of Mars as seen from Durban and Pretoria are shown in the diagram. The occultation on 31 January is visible from Central America and the southwestern United States.

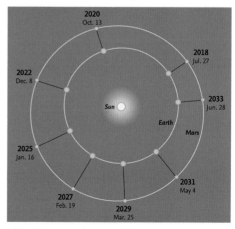

The oppositions of Mars between 2018 and 2033. As the illustration shows, there is no opposition in 2023.

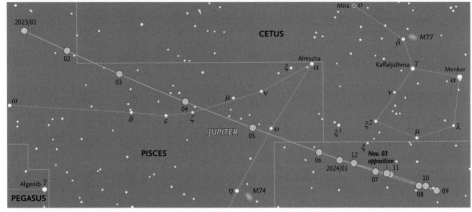

The path of Jupiter in 2023. Jupiter comes to opposition on November 3. Background stars are shown down to magnitude 6.5. South is up on all maps.

Jupiter and Saturn

IIn 2023, **Jupiter** spends the early part of the year in **Pisces**, starting in the west, then clips the top of **Cetus** and moves into **Aries** in mid-May. It begins retrograde motion in early September, and comes to opposition on November 3 at mag. -2.9. It continues retrograding to the end of the year, when it is still in Aries. **Saturn** starts the year in **Capricornus**, moving into **Aquarius** in February. It begins its retrograde motion in early July, and comes to opposition on August 27 at mag. 0.4. It continues retrograde motion until late October, reverting to direct motion and finishing the year in the western side of Aquarius.

Jupiter's four large satellites are readily visible in binoculars. Not all four are visible all the time, sometimes hidden behind the planet or invisible in front of it. **Io**, the closest to Jupiter, orbits in just under 1.8 days, and **Callisto**, the farthest away, takes about 16.7 days. In between are **Europa** (c. 3.6 days) and **Ganymede**, the largest, (c. 7.1 days). The diagram (below) shows the satellites' motions around the time of opposition.

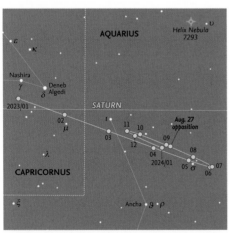

The path of Saturn in 2023. Saturn comes to opposition on August 27. Background stars are shown down to magnitude 6.5.

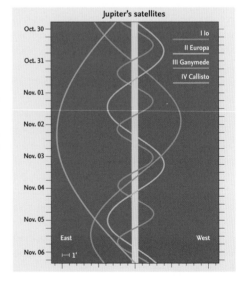

Uranus and Neptune

Uranus is in **Aries** for the whole of 2023. Initially retrograding, it reverts to direct motion in late January, then begins retrograde motion in early September. It reaches opposition (at mag. 5.6) on November 13 (at New Moon, so observation should be feasible). It is still retrograding at the end of the year, when it is mag. 5.6.

Neptune begins the year in **Aquarius** but moves into **Pisces** in March. It begins retrograde motion in mid-July and comes to opposition at mag. 7.8 on September 19. It moves back into Aquarius at the beginning of December, and resumes direct motion in late December when it is mag. 7.9.

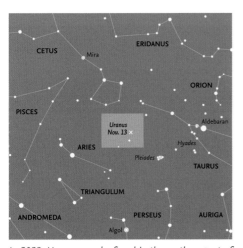

In 2023, Uranus may be found in the southern part of the constellation of Aries.

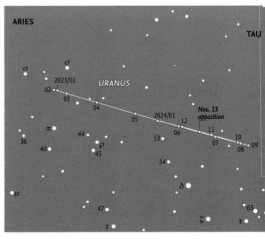

The path of Uranus in 2023. Uranus comes to opposition on November 13. All stars brighter than magnitude 7.5 are shown.

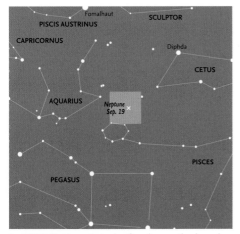

In 2023, Neptune is to be found near the boundary between Aquarius and Pisces. It comes to opposition on September 19.

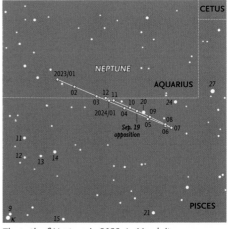

The path of Neptune in 2023. In March it moves across the boundary between the constellations of Aquarius and Pisces. All stars down to magnitude 8.5 are shown.

Minor Planets in 2023

Several minor planets rise above magnitude 9 in 2023, and four come to opposition above mag. 8.5. These are *(2) Pallas* in *Canis Major* on January 8 when it is mag. 7.7; *(1) Ceres* in *Coma Berenices* on March 21 at mag. 6.9; *(8) Flora* in *Aquarius* on August 27 (mag. 8.3) and *(4) Vesta* in the top of *Orion* at mag. 6.4 on December 21. Both (4) Vesta and the dwarf planet (1) Ceres are bright (above mag. 9.0) the whole year.

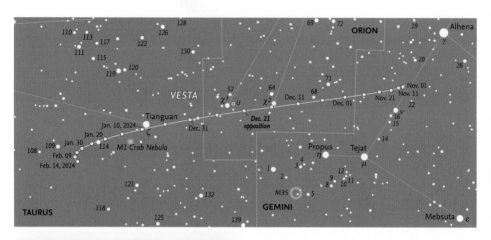

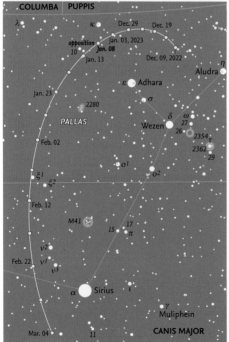

(Top) *The path of the minor planet (4) Vesta around its opposition on December 21 (mag. 6.4). The map also shows the path up to 14 February 2024.*

(Left) *The path of the minor planet (2) Pallas around its opposition on January 8 (mag. 7.7). The map also shows the path for December 2022.*

Both maps show background stars down to magnitude 8.0. South is up.

The four small maps at the bottom of these pages show the areas covered by the larger maps in a lighter blue. A white cross marks the position of the minor planet at the day of its opposition.

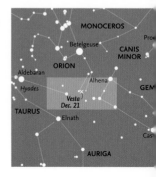

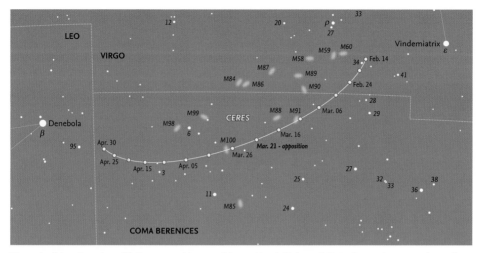

The path of the minor planet (1) Ceres around its opposition on March 21 (mag. 6.9). Background stars are shown down to magnitude 8.0. South is up.

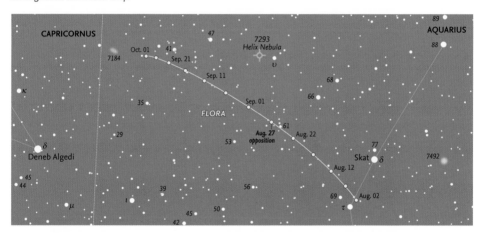

The path of the minor planet (8) Flora around its opposition on August 27 (mag. 8.3). Background stars are shown down to magnitude 9.0. South is up.

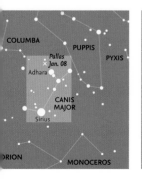
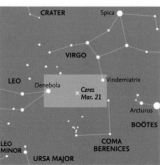
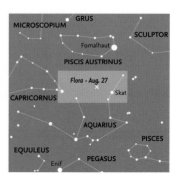

Comets in 2023

Although comets may occasionally become very striking objects in the sky, as was the case with Comet C/2020 F3 NEOWISE in 2020, their occurrence and particularly the existence or length of any tail and their overall magnitude are notoriously difficult to predict. Naturally, it is only possible to predict the return of periodic comets (whose names have the prefix 'P'). Many comets appear unexpectedly (these have names with the prefix 'C'). Bright, readily visible comets such as C/1995 Y1 Hyakutake, C/1995 O1 Hale-Bopp, C/2006 P1 McNaught or C/2020 F3 NEOWISE are rare. Most periodic comets are faint and only a very small number ever become bright enough to be easily visible with the naked eye or with binoculars.

The comets most likely to become visible in late 2023 are 103P/Hartley-2 (August to December) and 62P/Tsuchinshan-1 (November and December). Comet 62P/Tsuchinshan will probably continue bright into early 2024. Comet 2P/Encke will be very bright between October and December, but too close to the Sun to be readily visible.

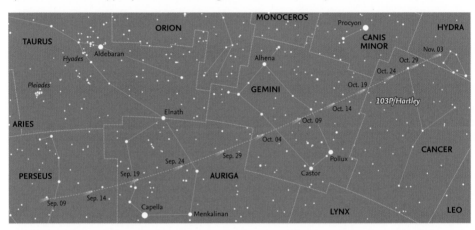

The path of Comet 103P/Hartley from 9 September to 3 November 2023. All stars down to magnitude 6.0 are shown. South is up.

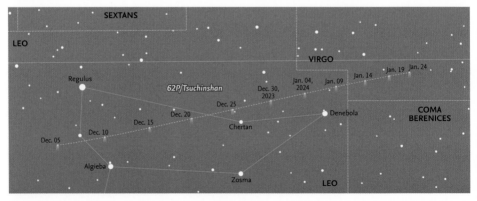

The path of comet 62P/Tsuchinshan in December 2023 and January 2024. Background stars are shown down to magnitude 6.5. South is up.

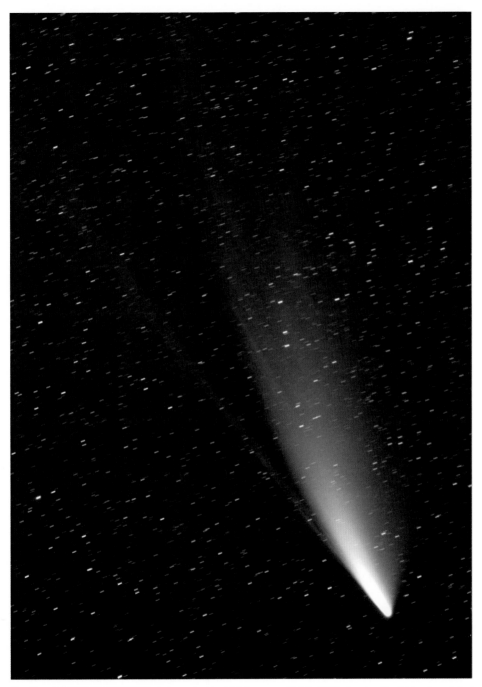

The Comet C/2020 F3 NEOWISE photographed by Nick James on 17 July 2020, showing the light-coloured dust tail, together with the blue ion tail.

Introduction to the Month-by-Month Guide

The monthly charts

The pages devoted to each month contain a pair of charts showing the appearance of the night sky, looking south and looking north. The charts (as with all the charts in this book) are drawn for the latitude of 35°S, so observers farther north will see slightly more of the sky on the northern horizon, and slightly less on the southern, with corresponding changes if they are farther south. The horizon areas are, of course, those most likely to be affected by poor observing conditions caused by haze, mist or smoke. In addition, stars close to the horizon are always dimmed by atmospheric absorption, so sometimes the faintest stars marked on the charts may not be visible.

The three times shown for each chart require a little explanation. The charts are drawn to show the appearance at 11 p.m. for the 1st of each month. The same appearance will apply an hour earlier (10 p.m.) on the 15th, and yet another hour earlier (9 p.m.) at the end of the month (shown as the 1st of the following month). Daylight Saving Time (DST) is not used in South Africa. In New Zealand, it applies from the last Sunday of September to the first Sunday of April. In those Australian states with DST, it runs from the first Sunday in October to the first Sunday in April. The appropriate times are shown on the monthly charts. Times of specific events are shown on the 24-hour clock of Universal Time (UT), used by astronomers worldwide, and corrections for the local time zone (and DST where employed) may be found from the details inside the front cover.

The charts may be used for earlier or later times during the night. To observe two hours earlier, use the charts for the preceding month; for two hours later, the charts for the next month.

Meteors

Details of specific meteor showers are given in the months when they come to maximum,

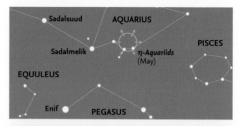

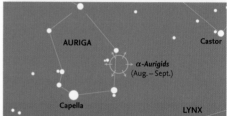

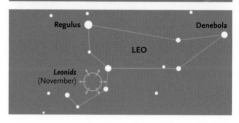

regardless of whether they begin or end in other months. Note that not all the respective radiants are marked on the charts for that particular month, because the radiants may be below the horizon, or lie in constellations that are not readily visible during the month of maximum. For this reason, special charts for the Eta (η) Aquariids (May), the Alpha (α) Aurigids (August and September) and the Leonids (November) are given here. As just explained, however, meteors from such showers may still be seen, because the most effective region for seeing meteors is some 40–45° away from the radiant, and that area of sky may well be above the horizon. A table of the best meteor showers visible during the year is also given here. The rates given are based on the properties of the meteor streams, and are those that an experienced observer might see under ideal conditions. Generally, the observed rates will be far less.

Shower	Dates of activity 2023	Date of maximum 2023	Possible hourly rate
α-Centaurids	January 31 to February 20	February 8	6
γ-Normids	February 25 to March 28	March 15	6
π-Puppids	April 15 to April 28	April 24	var.
η-Aquariids	April 19 to May 28	May 6	50
α-Capricornids	July 3 to August 15	July 30	5
Southern δ-Aquariids	July 12 to August 23	July 30	25
Piscis Austrinids	July 15 to August 10	July 28	5
Perseids	July 17 to August 24	August 13	100
α-Aurigids	August 28 to September 5	September 1	6
Southern Taurids	September 10 to November 20	October 10	5
Orionids	October 2 to November 7	October 21	20
Northern Taurids	October 20 to December 10	November 12	5
Leonids	November 6 to November 30	November 18	10
Phoenicids	November 28 to December 9	December 2	var.
Puppid Velids	December 1 to December 15	December 7	10
Geminids	December 4 to December 20	December 15	150

Meteors that are brighter than magnitude -4 (approximately the maximum magnitude reached by Venus) are known as *fireballs* or *bolides.*

The photographs

As an aid to identification – especially as some people find it difficult to relate charts to the actual stars they see in the sky – one or more photographs of constellations visible in certain specific months are included. It should be noted, however, that because of the limitations of the photographic and printing processes, and the differences between the sensitivity of different individuals to faint starlight (especially in their ability to detect different colours), and the degree to which they have become adapted to the dark, the apparent brightness of stars in the photographs will not necessarily precisely match that seen by any one observer.

The Moon calendar

The Moon calendar is largely self-explanatory. It shows the phase of the Moon for every day of the month, with the exact times (in Universal Time) of New Moon, First Quarter, Full Moon and Last Quarter. Because the times are calculated from the Moon's actual orbital parameters, some of the times shown will, naturally, fall during daylight, but any difference is too small to affect the appearance of the Moon on that date. Also shown is the *age* of the Moon (the day in the *lunation*), beginning at New Moon, which may be used to determine the best time for observation of specific lunar features.

The Moon

The section on the Moon includes details of any lunar or solar eclipses that may occur during the month (visible from anywhere on Earth). Similar information is given about any important occultations. Mainly, however, this section summarizes when the Moon passes close to planets or the five prominent stars close to the ecliptic. The dates when the Moon is closest to the Earth (at *perigee*) and farthest from it (at *apogee*) are shown in the monthly calendars, and only mentioned here when they are particularly significant, such as the nearest and farthest during the year

The planets and minor planets

Brief details are given of the location, movement and brightness of the planets from Mercury to Saturn throughout the month. None of the planets can, of course, be seen when they are close to the Sun, so such periods are generally noted. All of the planets may sometimes lie on the opposite side of the Sun to the Earth (at superior conjunction), but in the case of the inferior planets, Mercury and Venus, they may also pass between the Earth and the Sun (at inferior conjunction)

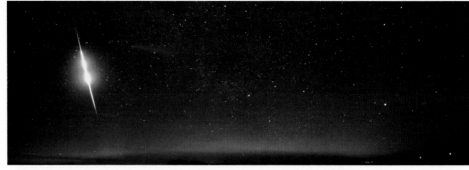

A fireball (with flares approximately as bright as the Full Moon), photographed against a weak auroral display by D. Buczynski from Tarbat Ness in Scotland on 22 January 2017.

and are normally invisible for a longer or shorter period of time. Those two planets are normally easiest to see around either eastern or western elongation, in the evening or morning sky, respectively. Not every elongation is favourable, so although every elongation is listed, only those where observing conditions are favourable are shown in the individual diagrams of events.

The dates at which the superior planets reverse their motion (from direct motion to retrograde, and retrograde to direct) and of opposition (when a planet generally reaches its maximum brightness) are given. Some planets, especially distant Saturn, may spend most or all of the year in a single constellation. Jupiter and Saturn are normally easiest to see around opposition, which occurs every year. Mars, by contrast, moves relatively rapidly against the background stars and in some years never comes to opposition. In 2020, Jupiter and Saturn are close together in the sky throughout the year, so a special chart is shown on page 24.

Uranus is normally magnitude 5.7–5.9, and thus at the limit of naked-eye visibility under exceptionally dark skies, but bright enough to be readily visible in binoculars. Because its orbital period is so long (over 84 years), Uranus moves only slowly along the ecliptic, and often remains within a single constellation for a whole year. The chart on page 25 shows its position during 2023.

Similar considerations apply to Neptune, although it is always fainter (generally magnitude 7.8–8.0), still visible in most binoculars. It takes about 164.8 years to complete one orbit of the Sun. As with Uranus, it frequently spends a complete year in one constellation. Its chart is also on page 25.

In any year, few minor planets ever become bright enough to be detectable in binoculars. Just one, (4) Vesta, on rare occasions brightens sufficiently for it to be visible to the naked eye. Our limit for visibility is magnitude 9.0 and details and charts are given for those objects that exceed that magnitude during the year, normally around opposition. To assist in recognition of a planet or minor planet as it moves against the background stars, the latter are shown to a fainter magnitude than the object at opposition. Minor-planet charts for 2023 are on pages 26 and 27.

The ecliptic charts

Although the ecliptic charts are primarily designed to show the positions and motions of the major planets, they also show the motion of the Sun during the month. The light-tinted area shows the area of the sky that is invisible during daylight, but the darker area gives an indication of which constellations are likely to be visible at some time of the night. The closer a planet is to the border between dark and light, the more difficult it will be to see in the twilight.

The monthly calendar

For each month, a calendar shows details of significant events, including when planets are close to one another in the sky, close to the Moon, or close to any one of five bright stars that are spaced along the ecliptic. The times shown are given in Universal Time (UT), always used by astronomers throughout the year, and which is identical to Greenwich Mean Time (GMT). So during the summer months, they do not show Summer Time, which will need to be taken for the observing location.

The diagrams of interesting events

Each month, a number of diagrams show the appearance of the sky when certain events take place. However, the exact positions of celestial objects and their separations greatly depend on the observer's position on Earth. When the Moon is one of the objects involved, because it is relatively close to Earth, there may be very significant changes from one location to another. Close approaches between planets or between a planet and a star are less affected by changes of location, which may thus be ignored.

The diagrams showing the appearance of the sky are drawn at latitude 35°S and longitude 150°E (approximately that of Sydney, Australia), so will be approximately correct for much of Australia. However, for an observer farther north (say, Brisbane or Darwin), a planet or star listed as being north of the Moon will appear even farther north, whereas one south of the Moon will appear closer to it – or may even be hidden (occulted) by it. For an observer at a latitude greater than 35°S (such as in New Zealand), there will be corresponding changes in the opposite direction: for a star or planet south of the Moon, the separation will increase, and for one north of the Moon, the separation will decrease. This is particularly important when occultations occur, which may be visible from one location, but not another. There are two occultations of Mars in 2023, one of which is visible from southern Africa.

Ideally, details should be calculated for each individual observer, but this is obviously impractical. In fact, positions and separations are actually calculated for a theoretical observer located at the centre of the Earth.

So the details given regarding the positions of the various bodies should be used as a guide to their location. A similar situation arises with the times that are shown. These are calculated according to certain technical criteria, which need not concern us here. However, they do not necessarily indicate the exact time when two bodies are closest together. Similarly, dates and times are given, even if they fall in daylight, when the objects are likely to be completely invisible. However, such times do give an indication that the objects concerned will be in the same general area of the sky during both the preceding and the following nights.

Data used in this Guide

The data given in this Guide, such as timings and distances between objects, have been computed with a program developed by the US Naval Observatory in Washington DC, widely regarded as the most accurate computation. As such, the data may differ slightly from information given elsewhere.

Key to the symbols used on the monthy star maps.

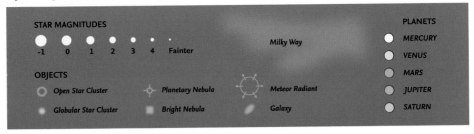

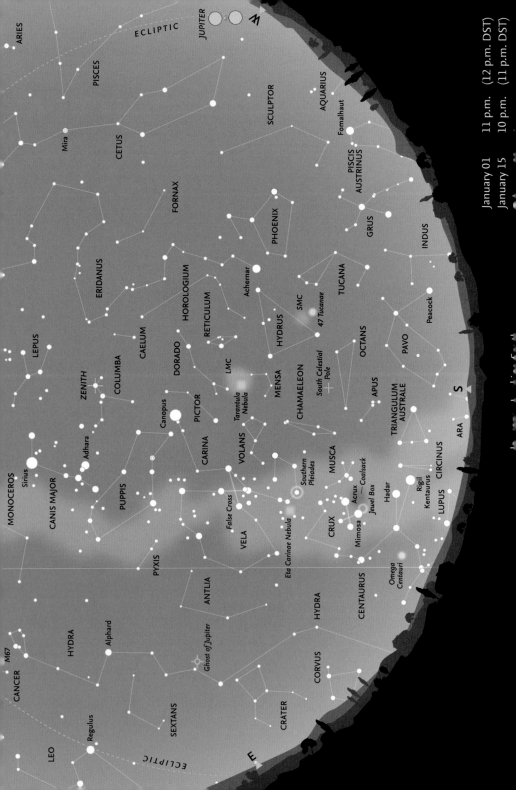

ARIES

ECLIPTIC

JUPITER

PISCES

SCULPTOR

AQUARIUS

Mira ⊛

CETUS

Fomalhaut

PISCIS
AUSTRINUS

FORNAX

PHOENIX

GRUS

INDUS

ERIDANUS

Achernar

HOROLOGIUM

RETICULUM

HYDRUS

SMC

TUCANA

LEPUS

CAELUM

LMC

47 Tucanae

Peacock

COLUMBA

DORADO

MENSA

OCTANS

PAVO

ZENITH

South Celestial
Pole

CHAMAELEON

Canopus

PICTOR

Tarantula
Nebula

APUS

Adhara

VOLANS

TRIANGULUM
AUSTRALE

Sirius

CARINA

Southern
Pleiades

MUSCA

ARA

MONOCEROS

PUPPIS

Acrux
— Coalsack

S

CANIS MAJOR

False Cross

Jewel Box

Hadar

Rigil
Kentaurus

CIRCINUS

VELA

CRUX

Mimosa

LUPUS

Eta Carinae Nebula

PYXIS

ANTLIA

Omega
Centauri

HYDRA

CENTAURUS

M67

Alphard

CANCER

HYDRA

Ghost of Jupiter

CORVUS

LEO

SEXTANS

CRATER

Regulus

ECLIPTIC

E

January 01 11 p.m. (12 p.m. DST)
January 15 10 p.m. (11 p.m. DST)

January – Looking South

Crux is now quite prominent as it rises in the east, and even **Rigil Kentaurus** and **Hadar** (α and β Centauri), although low, are becoming easier to see. The whole of **Carina** is visible, with both the **Eta Carinae Nebula** and the **Southern Pleiades** between Crux and the **False Cross** of stars from the constellations of Carina and **Vela**. **Canopus** (α Carinae) is high in the south, roughly three-quarters of the way from the horizon to the zenith. Also in the south, but slightly lower, is the **Large Magellanic Cloud** (LMC). **Achernar** (α Eridani) is prominent between the constellations of **Hydrus** and **Phoenix**. The **Small Magellanic Cloud** (SMC) and **47 Tucanae** are well-placed for observation alongside Hydrus. **Fomalhaut** (α Piscis Austrini) may be glimpsed low on the horizon towards the west as may **Peacock** (α Pavonis) farther towards the south. The whole of **Pavo** and **Grus** is visible, together with most of **Indus**.

Meteors

The year does not start well for southern meteor observers. The **α-Centaurids** (active late January to February) have low maximum rates, as do the **γ-Normids** (February to March). Only with the **π-Puppids** (maximum April 23–24) is there any likelihood of a higher rate of about 40, although even that is rare. The **Southern δ-Aquarids** (maximum July 30) have a rate of about 20 per hour, as do the **Orionids** (October 21). Only with the **Leonids** (maximum November 17–18) is there a rate of 10 per hour. The **Phoenicids** in November–December (maximum December 2) are unpredictable, but may show strong activity. The last major shower of the year, the **Geminids** (maximum December 15) finally achieves a rate of about 120 or even 150 per hour.

Part of the southern Milky Way. The two bright stars near the bottom of the image are α and β Centauri. The constellation of Crux is near the centre. The Eta Carinae Nebula and the Southern Pleiades are near the top-right corner.

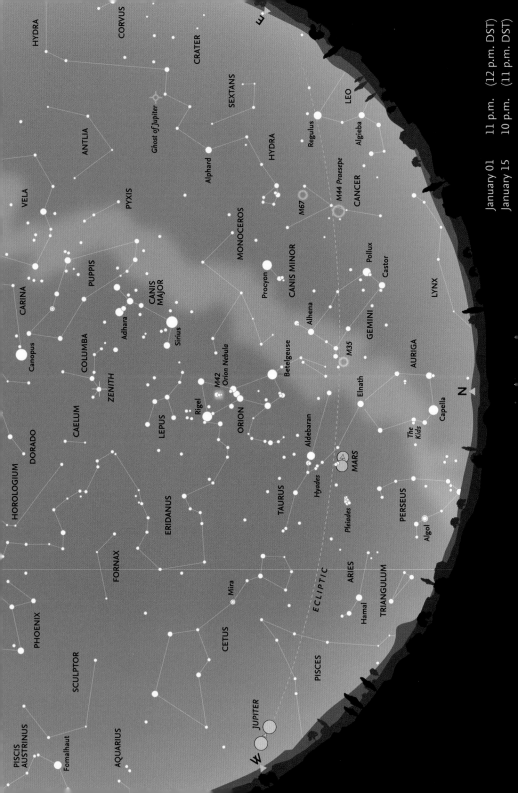

E

January 01   11 p.m.   (12 p.m. DST)
January 15   10 p.m.   (11 p.m. DST)

January – Looking North

The northern sky is dominated by **Orion**, prominent during the summer months and visible at some time during the night. It is highly distinctive, with a line of three stars that form the 'Belt'. To most observers, the bright star **Betelgeuse** (α Orionis), shows a reddish tinge, in contrast to the brilliant bluish-white **Rigel** (β Orionis). The three stars of the belt lie directly south of the celestial equator. A vertical line of three 'stars' forms the 'Sword' that hangs south of the Belt. With good viewing, the central 'star' appears as a hazy spot, even to the naked eye, and is actually the **Orion Nebula**. Binoculars reveal the four stars of the Trapezium, which illuminate the nebula.

Orion's Belt points down to the northwest towards **Taurus** (the Bull) and orange-tinted **Aldebaran** (α Tauri). Close to Aldebaran, there is a conspicuous 'V' of stars, called the **Hyades** cluster. (Despite appearances, Aldebaran is not part of the cluster.) Farther along, the same line from Orion passes above a bright cluster of stars, the **Pleiades**, or Seven Sisters. Even the smallest pair of binoculars reveals this as a beautiful group of bluish-white stars. The two most conspicuous of the other stars in Taurus lie directly below Orion, and form an elongated triangle with Aldebaran. The northernmost, **Elnath** (β Tauri), was once considered to be part of the constellation of **Auriga**.

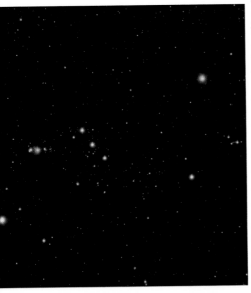

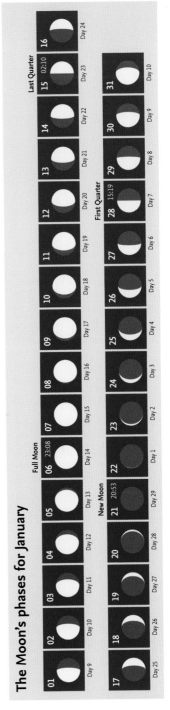

The constellation of Orion dominates the northern sky during this period of the year, and is a useful starting point for recognizing other constellations in the area (see page 36). Here, orange Betelgeuse, blue-white Rigel and the pinkish Orion Nebula are prominent (south is up).

The Moon's phases for January

				Full Moon					
01	02	03	04	05	06 23:08	07	08	09	10
Day 25	Day 26	Day 27	Day 28	Day 29	Day 1	Day 2	Day 3	Day 4	Day 5

		New Moon				First Quarter			Last Quarter
17	18	19	20	21 20:53	22	23	24		15 02:10	16
Day 9	Day 10	Day 11	Day 12	Day 13	Day 14	Day 15	Day 16	Day 17	Day 23	Day 24

| 11 | 12 | 13 | 14 | 15 02:10 | 16 |
| Day 18 | Day 19 | Day 20 | Day 21 | Day 22 | Day 23 |

| 25 | 26 | 27 | 28 15:19 | 29 | 30 | 31 |
| Day 17 | Day 18 | Day 19 | Day 20 | Day 7 | Day 8 | Day 9 | Day 10 |

January – Moon and Planets

The Earth

The Earth reaches perihelion (the closest point to the Sun in its yearly orbit) on January 4 at 16:17 Universal Time, when its distance is 0.983295578 AU (147,098,855 km).

The Moon

On January 1, the Moon passes 0.7° north of *Uranus* (mag. 5.7). On January 3 it occults *Mars* in *Taurus* (this event is visible only from the Indian Ocean and southern Africa), and the next day is 8.1° north of *Aldebaran*. On January 7, it is 1.9° south of *Pollux* in *Gemini*. On January 10, it passes 4.6° north of *Regulus*, between it and *Algieba*. By January 14, a day before Last Quarter it will be 3.8° north of *Spica* in *Virgo*. As a waning crescent it will be 2.1° north of *Antares* on January 18. On January 20, one day before New

Moon, it is 6.9° south of *Mercury* (mag. -0.6) just past inferior conjunction and invisible in the evening sky. On January 23, the thin waxing crescent will be 3.8° south of *Saturn* (mag. 0.8), and a little later, 3.5° south of *Venus*, much brighter at mag. -3.9. On January 31 it occults *Mars* again, this time visible from Central America and the southwestern United States.

The planets

Mercury passes inferior conjunction on January 7. Although it is bright (mag. -3.9), *Venus* is close to the Sun in twilight. *Mars* is in Taurus, slowly fading from mag. -1.2 to -0.3 over the month. It is occulted by the Moon on January 3 and 31. *Jupiter* (average mag. -2.) is in *Pisces* and *Saturn* (mag. 0.8) in *Capricornus*, close to the Sun in evening twilight. *Uranus*, initially retrograding until January 23, is mag. 5.7 in *Aries*. *Neptune* is mag. 7.9 in *Aquarius*. On January 8, the minor planet (2) *Pallas* is at opposition in *Canis Major* at mag. 7.7 (see the chart on page 26). On January 26, minor planet (6) *Hebe* is at opposition in *Hydra* at mag. 8.8.

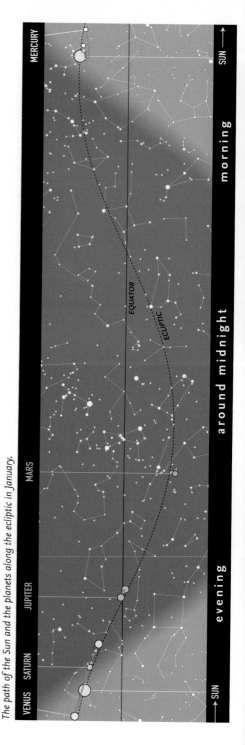

The path of the Sun and the planets along the ecliptic in January.

Calendar for January

01	22:16	Uranus 0.7°S of the Moon
03	19:38	Mars occulted by the Moon
04	00:53	Aldebaran 8.1°S of the Moon
04	16:17	Earth at perihelion (0.98329578 AU)
06	23:08	Full Moon
07	12:57	Mercury at inferior conjunction
07	14:18	Pollux 1.9°N of the Moon
08	09:19	Moon at apogee = 406,458 km
08	19:02	Minor planet (2) Pallas at opposition (mag. 7.1)
10	12:17	Regulus 4.6°S of the Moon
14	22:37	Spica 3.8°S of the Moon
15	02:10	Last Quarter
18	10:04	Antares 2.1°S of the Moon
20	07:49	Mercury 6.9°N of the Moon
21	20:53	New Moon
21	20:57	Moon at perigee = 356,569 km
22	20:00 *	Saturn (mag. 0.8) 0.4°N of Venus (mag. -3.9)
23	07:21	Saturn 3.8°N of the Moon
23	08:18	Venus 3.5°N of the Moon
25	05:55	Neptune 2.7°N of the Moon
26	02:03	Jupiter 1.8°N of the Moon
26	09:17	Minor planet (6) Hebe at opposition (mag. 8.8)
28	15:19	First Quarter
29	04:09	Uranus 1.0°S of the Moon
30	05:54	Mercury at greatest elongation (25.0°W, mag. -0.1)
31–Feb.20		α–Centaurid meteor shower
31	04:25	Mars occulted by the Moon
31	06:45	Aldebaran 8.3°S of the Moon

* These objects are close together for an extended period around this time.

Evening 11:30 p.m. (DST)

January 7 • The Full Moon is close to Pollux, with Castor nearby.

Morning 5 a.m. (DST)

January 18–19 • The Moon passes Antares, between it and Sabik (η Oph).

Evening 8:45 p.m. (DST)

January 23 • The Moon is close to the western horizon with Venus and the much fainter Saturn (mag. 1.2).

Evening 9 p.m. (DST)

January 26 • The waxing crescent Moon is near Jupiter, in the west-northwest.

Evening 9 p.m. (DST)

January 30–31 • The Moon passes the Pleiades, on January 30. One day later it is near Mars, due north, with the orange coloured Aldebaran just eight degrees above it.

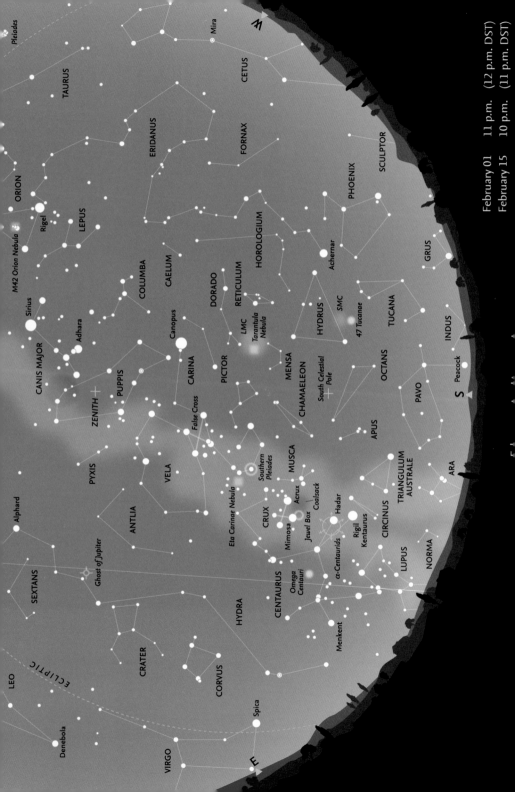

Pleiades

TAURUS

ORION

M42 Orion Nebula

Rigel

CANIS MAJOR

Sirius

Adhara

LEPUS

ERIDANUS

COLUMBA

CAELUM

Canopus

CARINA

PICTOR

DORADO

RETICULUM

HOROLOGIUM

LMC

Tarantula Nebula

MENSA

CHAMAELEON

South Celestial Pole

OCTANS

HYDRUS

SMC

47 Tucanae

TUCANA

PAVO

APUS

CETUS

FORNAX

PHOENIX

SCULPTOR

Achernar

GRUS

INDUS

Peacock

S

ARA

TRIANGULUM AUSTRALE

CIRCINUS

NORMA

LUPUS

MUSCA

Acrux

Coalsack

Jewel Box

Mimosa

CRUX

Hadar

α-Centaurids

Rigil Kentaurus

Omega Centauri

CENTAURUS

Menkent

Southern Pleiades

Eta Carinae Nebula

VELA

PYXIS

ANTLIA

Ghost of Jupiter

Alphard

SEXTANS

HYDRA

CRATER

CORVUS

Spica

VIRGO

Denebola

LEO

ECLIPTIC

ZENITH

PUPPIS

False Cross

Mira

February 01   11 p.m.   (12 p.m. DST)
February 15   10 p.m.   (11 p.m. DST)

W

E

N

February – Looking South

The whole of **Centaurus** is now clear of the horizon in the southeast, where, below it, the constellation of **Lupus** is beginning to be visible. **Triangulum Australe** is now higher in the sky and easier to see. Both **Crux** and the **False Cross** are also higher and more visible. The **Large Magellanic Cloud** (LMC) and brilliant **Canopus** are now almost due south, with the constellation of **Puppis** at the zenith. Both **Hydrus** and the **Small Magellanic Cloud** (SMC) are lower in the sky, as is **Phoenix**, which is much closer to the horizon in the west. Next to it, the constellation of **Tucana** is also lower although the globular cluster **47 Tucanae** remains readily visible as does **Achernar** (α Eridani). The constellation of **Grus** and bright **Fomalhaut** (α Piscis Austrini) have disappeared below the horizon. **Peacock** (α Pavonis) is skimming the horizon in the south and is not easily seen at any time in the night.

Meteors

The **Centaurid** shower (which actually consists of two separate streams: the **α- and β-Centaurids**, with both radiants lying near α and β Centauri (Rigil Kentaurus and Hadar, respectively) continues in February, reaching a low maximum (around 5 meteors per hour) on February 8, when the Moon is waning gibbous, just after Full Moon. Another weak shower, the γ-**Normids**, begins to be active in late February (February 25), but the meteors are difficult to differentiate from sporadics. It reaches its weak (but sharp) maximum on March 14–15.

Parts of Carina and Vela. The 'False Cross' is indicated with blue lines. The red blurry spot near the bottom-left corner is the Eta (η) Carinae Nebula. To the right, and a little lower, is an open cluster that surrounds Theta (θ) Carinae. This cluster (IC 2602) is also known as the Southern Pleiades.

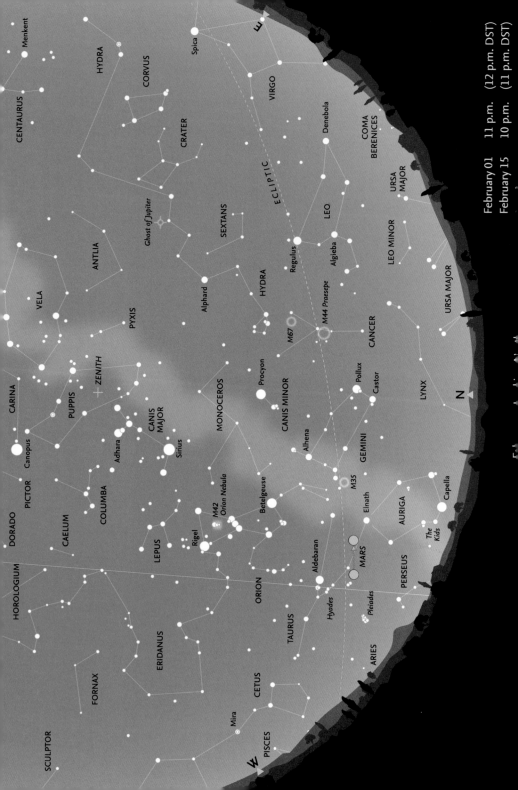

February – Looking North

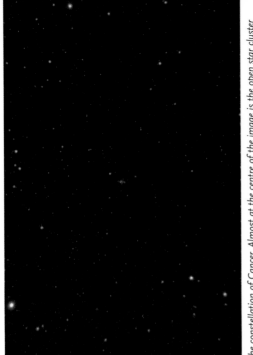

The constellation of **Gemini**, with the pair of stars **Castor** and **Pollux** (α and β Geminorum) is now due north. Pollux is higher in the sky (farther towards the zenith), and above it is **Procyon** (α Canis Minoris), halfway to the zenith. Still higher is **Sirius**, the brightest star in the sky and the constellation of **Canis Major**. The faint constellation of **Cancer**, with its most noticeable feature, the cluster **Praesepe**, lies to the east of Gemini. Above it, and directly east of Procyon, is the distinctive asterism forming the head of **Hydra**, the whole of which constellation is now visible stretching across the sky towards the east. The constellation of **Taurus**, with orange **Aldebaran** (α Tauri) is still clearly seen in the west. By contrast, **Auriga** is much lower towards the horizon and brilliant **Capella** (α Aurigae) is extremely low and visible only early in the night. The faintest stretch of the Milky Way runs from Auriga in the northwest up towards the zenith, passing through Gemini and the indistinct constellation of **Monoceros**. In the northeast, the zodiacal constellation of **Leo** and bright **Regulus** (α Leonis) are clearly seen.

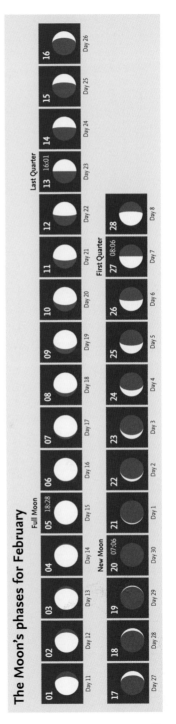

The constellation of Cancer. Almost at the centre of the image is the open star cluster M44 (Praesepe), sometimes called the 'Beehive'. The faint constellation of Cancer is surrounded by several bright stars; Castor and Pollux, near the bottom, Procyon near the top and Regulus with Algieba close to the right edge of the image (south is up).

The Moon's phases for February

				Full Moon					
01	02	03	04	05 18:28	06	07	08	09	10
Day 11	Day 12	Day 13	Day 14	Day 15	Day 16	Day 17	Day 18	Day 19	Day 20

						Last Quarter		
11	12	13 16:01	14	15	16			
Day 21	Day 22	Day 23	Day 24	Day 25	Day 26			

	New Moon							First Quarter	
17	18	19	20 07:06	21	22	23	24	25	26	27 08:06	28
Day 27	Day 28	Day 29	Day 30	Day 1	Day 2	Day 3	Day 4	Day 5	Day 6	Day 7	Day 8

February – Moon and Planets

The Moon

On February 3, two days before Full Moon, the Moon passes 1.9° south of *Pollux* in *Gemini*. One day after Full Moon, it is 4.5° north of *Regulus* in *Leo*. On February 11, it is 3.5° north of *Spica* in *Virgo*. On February 14, one day after Last Quarter, the Moon is 1.8° north of *Antares* in *Scorpius*. On February 19, one day before New Moon, it is 3.7° south of *Saturn*, but this will be lost in twilight. The same problem will apply on February 21, when the Moon is 2.5° south of *Neptune*. The next day, the Moon is 2.1° south of Venus, which is mag. -3.9, so will be visible in the evening twilight. Later the same day, the Moon is 1.2° south of *Jupiter* (mag. -2.1) in *Pisces*. On February 28, the Moon is 1.1° north of *Mars*, in *Taurus*.

Occultations

In 2023 there are no lunar occultations of the five brightest stars near the ecliptic: *Aldebaran, Antares, Pollux, Regulus* and *Spica*, nor of stars in the *Pleiades*. There are two full occultations of *Mars*, the first on January 3, visible only from the Indian Ocean and southern Africa, and the second, on January 31, visible from Central America and the southwestern United States. The start of an occultation of *Venus* on March 24 may be visible from part of southeastern Asia. There are no major occultations of *Jupiter* and none of *Saturn*.

The planets

Mercury is invisible in twilight. *Venus* is bright (mag. -3.9) and will become visible in the evening sky towards the end of the month. *Mars* fades from mag. -0.3 to mag. 0.4 over the month and is in *Taurus*. *Jupiter* is mag. -2.1 and is in *Pisces*, but *Saturn* is much closer to the Sun, and will be lost in twilight. *Uranus* remains in *Aries* at mag. 5.8, and *Neptune* in *Pisces* at mag. 7.9 to 8.0, close to evening twilight.

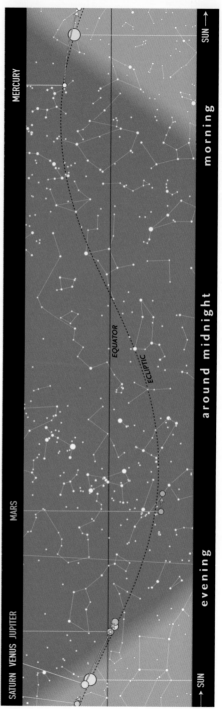

The path of the Sun and the planets along the ecliptic in February.

Calendar for February

03	20:25	Pollux 1.9°N of the Moon
04	08:55	Moon at apogee = 406,476 km
05	18:28	Full Moon
06	18:22	Regulus 4.5°S of the Moon
08		α-Centaurid meteor shower maximum
11	05:02	Spica 3.5°S of the Moon
14	16:01	Last Quarter
14	18:43	Antares 1.8°S of the Moon
15	12:00 *	Neptune (mag. 8.0) 0.01°N of Venus (mag. -3.9)
16	16:48	Saturn at superior conjunction
18	20:52	Mercury 3.6°N of the Moon
19	09:06	Moon at perigee = 358,267 km
19	23:58	Saturn 3.7°N of the Moon
20	07:06	New Moon
21	18:16	Neptune 2.5°N of the Moon
22	07:55	Venus 2.1°N of the Moon
22	22:00	Jupiter 1.2°N of the Moon
25–Mar.28		γ-Normid meteor shower
25	13:05	Uranus 1.3°S of the Moon
27	08:06	First Quarter
27	13:39	Aldebaran 8.5°S of the Moon
28	04:32	Mars 1.1°S of the Moon

These objects are close together for an extended period around this time.

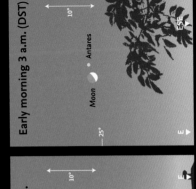

Evening 10 p.m. (DST)

February 3 • The Moon forms a nice, almost isosceles, triangle with Pollux and Castor.

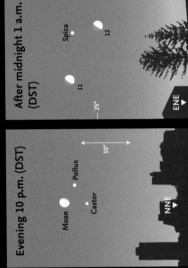

After midnight 1 a.m. (DST)

February 11–12 • The Moon passes below Spica, in the constellation of Virgo.

Evening 8:15 p.m. (DST)

February 22–23 • In the evening twilight, the narrow crescent Moon passes Venus and Jupiter.

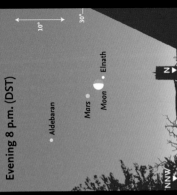

Early morning 3 a.m. (DST)

February 15 • High in the east, the Moon is close to Antares.

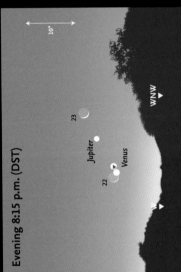

Evening 8 p.m. (DST)

February 28 • In the north, the Moon is lining up with Elnath (β Tau), Mars and Aldebaran.

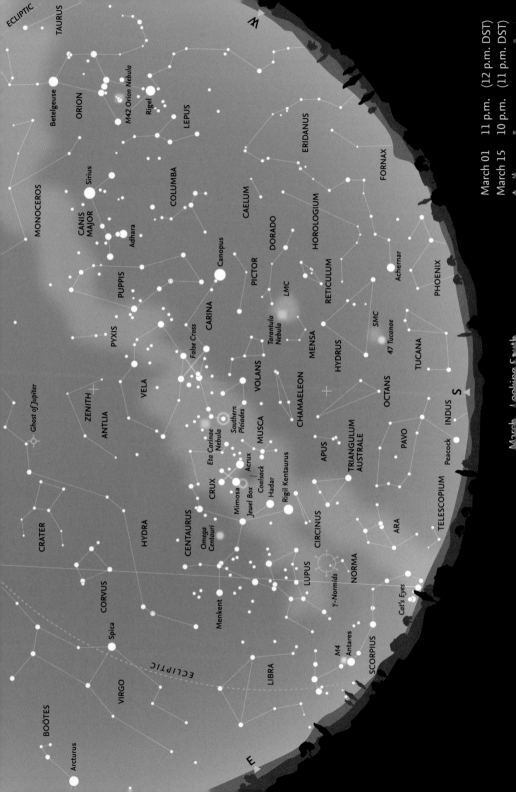

March — Looking South

March 01   11 p.m.   (12 p.m. DST)
March 15   10 p.m.   (11 p.m. DST)

The constellation of Vela may be found in the lower part of the image, as well as the False Cross, consisting of two stars that belong to Vela, while the two stars near the bottom are part of Carina. The constellation of Pyxis is near the top (north is up).

March – Looking South

The Sun crosses the celestial equator on Monday, March 20, at the equinox, when day and night are of almost equal length, and the southern season of autumn is considered to begin. (The hours of daylight and darkness change most rapidly around the equinoxes in March and September.)

Scorpius is beginning to become visible in the eastern sky, including the 'Cat's Eyes', the pair of stars **Shaula** and **Lesath** (λ and υ Scorpii, respectively) at the end of the 'sting'. Above Scorpius, the whole of the constellation of **Lupus** is now easy to see. The magnificent globular cluster of **Omega Centauri** is now readily visible, north-east of **Crux**. The **Coalsack** and the denser region of the Milky Way in **Carina**, together with the **Eta Carinae Nebula** and the **Southern Pleiades** are well placed for observation. The **False Cross** on the **Carina/Vela** border is now high in the sky, between the South Celestial Pole and the zenith. Brilliant **Canopus** (α Carinae) is only slightly lower towards the west, above the **Large Magellanic Cloud** (LMC) and the striking **Tarantula Nebula**. **Achernar** (α Eridani), the **Small Magellanic Cloud** (SMC) and **47 Tucanae** are considerably lower, but still clear of the horizon. **Peacock** (α Pavonis) remains low, skimming the horizon, just east of south. **Orion** is now visibly getting lower in the west, and is being followed by **Sirius** and **Canis Major**.

Meteors

The only significant meteor shower in March is the *γ*-**Normids**, which have a low rate, and are thus difficult to differentiate from sporadics. However, they exhibit a very sharp peak a day or so on either side of maximum on March 14–15, just before and at Last Quarter, so conditions will be moderately favourable. The faint constellation of **Norma** rises early in the night, but most meteors are likely to be seen (away from the radiant) in the hours after midnight.

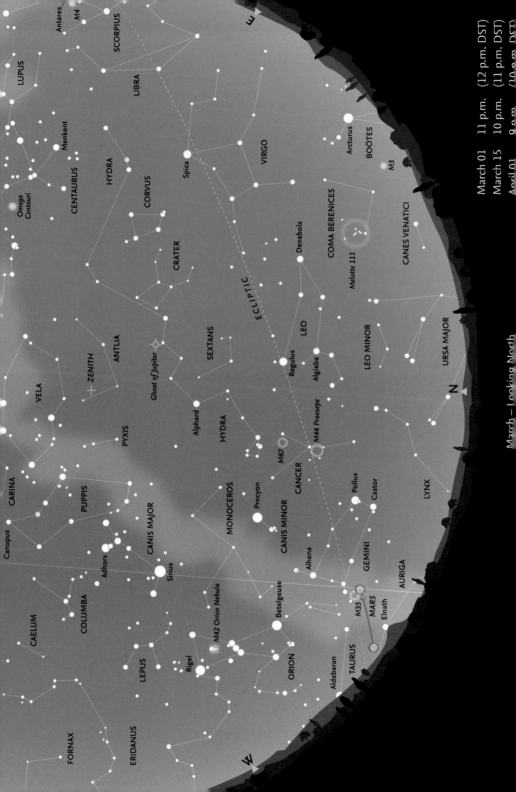

March – Looking North

March 01	11 p.m.	(12 p.m. DST)
March 15	10 p.m.	(11 p.m. DST)
April 01	9 p.m.	(10 p.m. DST)

March – Looking North

Almost due north is the constellation of *Leo*, with the 'backward question mark' (or 'Sickle') of bright stars forming the head of the mythological lion. **Regulus** (α Leonis) – the 'dot' of the 'question mark' or the handle of the sickle and the brightest star in Leo – lies very close to the ecliptic and is one of the few first-magnitude stars that may be occulted by the Moon, although none occur in 2023. The Moon often passes between Regulus and **Algieba** (γ Leonis). To the west lies the faint constellation of *Cancer*, with the open cluster M44, or **Praesepe**. The constellations of *Gemini* and *Orion* are now getting low in the west, but above them, both **Procyon** in *Canis Minor* and the constellation of *Canis Major* remain clear to see. The whole of **Hydra** (the largest constellation) now sprawls right across the southern and eastern skies, and the three small constellations of *Sextans*, *Crater* and *Corvus* are readily visible. The unremarkable constellation of *Antlia* is at the zenith. In the east, *Arcturus* (α Boötis) – the brightest star in the northern hemisphere of the sky – is becoming visible, and climbs higher during the night.

The Moon's phases for March

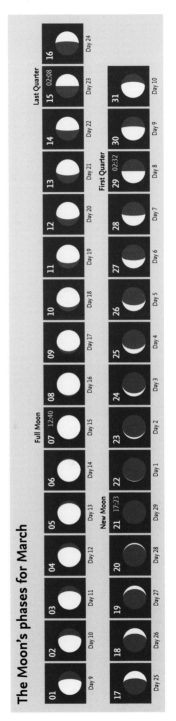

The distinctive constellation of Leo, with Regulus and 'The Sickle' on the west. Algieba (γ Leonis), north of Regulus, appearing double, is a multiple system of four stars (south is up).

March – Moon and Planets

The Moon

On March 3, the waxing gibbous Moon is 1.7° south of *Pollux* (mag. 1.1) the brightest star in *Gemini*. On March 6, one day before Full Moon, it is 4.5° north of *Regulus* (mag. 1.4) in *Leo*. By March 10 it is 3.4° north of *Spica* and by March 14, one day before Last Quarter, it is 1.6° north of *Antares* in *Scorpius*. On March 19, the waning crescent is 3.6° south of *Saturn* (mag. 0.8). At New Moon, on March 21, it is 2.4° south of faint *Neptune* (mag. 8.0). The next day in twilight, it passes south of *Mercury*, and then 0.5° south of *Jupiter* (mag. -2.1). On March 24, the Moon occults *Venus*, partly visible from southeast Asia. On March 26, the waxing crescent is 8.7° north of *Aldebaran* in *Taurus*. It passes 2.3° north of *Mars* on March 28 and on March 30 is again 1.6° south of *Pollux.*

The Planets

Mercury is too close to the Sun to be seen. It reaches superior conjunction, on the far side of the Sun, on March 17. *Venus*, in the evening sky, is very bright (mag. -3.9 to -4.0), but too close to the Sun to be easily visible. *Mars* is initially mag. 0.4 in *Taurus*, but moves into *Gemini* and fades to mag. 1.0. *Jupiter* is in *Pisces*, but is too close to the Sun to be readily visible this month. *Saturn* is in *Aquarius* and too deep in the morning twilight to be seen. *Uranus* is in *Aries* at mag. 5.8 and *Neptune* (mag. 8.0) is in *Pisces* and comes to conjunction on March 15.

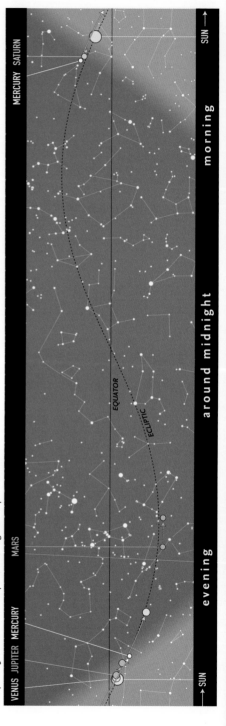

The path of the Sun and the planets along the ecliptic in March.

Calendar for March

02	10:00 *	Saturn (mag. 0.9) 0.9°N of Mercury (mag. -0.6)
02	11:00 *	Jupiter (mag. -2.1) 0.5°S of Venus (mag. -3.9)
03	02:48	Pollux 1.7°N of the Moon
03	15:00	Moon at apogee = 405,889 km
06	00:46	Regulus 4.5°S of the Moon
07	12:40	Full Moon
10	10:45	Spica 3.4°S of the Moon
14	00:56	Antares 1.6°S of the Moon
15		γ-Normid meteor shower maximum
15	02:08	Last Quarter
15	23:39	Neptune at superior conjunction
16	15:00 *	Neptune (mag. 8.0) 0.4°N of Mercury (mag. -1.8)
17	10:45	Mercury at superior conjunction
19	15:12	Moon at perigee = 362,697 km
19	15:22	Saturn 3.6°N of the Moon
20	21:24	Autumnal equinox
21	06:47	Neptune 2.4°N of the Moon
21	07:41	Dwarf planet Ceres at opposition (mag. 6.9)
21	17:23	New Moon
22	00:10	Mercury 1.8°N of the Moon
22	19:56	Jupiter 0.5°N of the Moon
24	10:26	Venus occulted by the Moon
25	00:39	Uranus 1.5°S of the Moon
26	01:00	European Daylight Saving Time (British Summer Time, BST) begins
26	22:05	Aldebaran 8.7°S of the Moon
28	13:16	Mars 2.3°S of the Moon
28	15:00 *	Mercury (mag. -1.4) 1.5°N of Jupiter (mag. -2.1)
29	02:32	First Quarter
30	10:02	Pollux 1.6°N of the Moon
31	06:00 *	Uranus (mag. 5.8) 1.3°S of Venus (mag. -4.0)
31	11:17	Moon at apogee = 404,919 km

*These objects are close together for an extended

Evening 8:15 p.m. (DST)

March 2 • Low on the western horizon, Jupiter and Venus are close together. Diphda (β Cet) is about twelve degrees higher.

Evening 7:30 p.m. (DST)

March 23–24 • After sunset on March 23, the narrow crescent Moon is near Jupiter, but may not

After midnight 0:30 a.m. (DST)

March 14–15 • The Moon passes below Antares. The name Antares means 'Rival of Mars', because of its similar colour. The Cat's Eyes (λ and υ Sco) are farther southeast.

Evening 8 p.m. (DST)

March 27–30 • The Moon passes Elnath (β Tau), Mars and Pollux. Castor and Alhena (γ Gem) are nearby.

MARCH 5

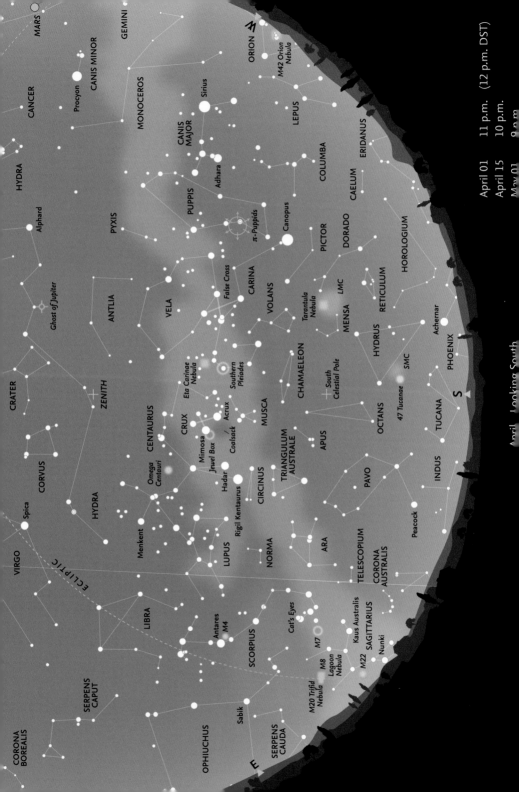

April, Looking South

April 01   11 p.m. (12 p.m. DST)
April 15   10 p.m.
May 01   9 p.m.

April – Looking South

Daylight Saving Time ends in both Australia and New Zealand on Sunday, April 2 with the arrival of autumn. **Crux** is now high in the south, with the two brightest stars of **Centaurus** (α and β Centauri, **Rigil Kentaurus** and **Hadar**, respectively) to its east. The magnificent globular cluster, **Omega Centauri**, is readily visible high in the sky. West of Crux, both the **Southern Pleiades** and the **Eta Carinae Nebula** are clearly seen, two-thirds of the way towards the zenith. Farther west, the **False Cross** is beginning to decline towards the horizon. **Canopus** (α Carinae) is even lower, and **Orion** has now disappeared below the horizon. **Canis Major** and brilliant **Sirius** are also descending in the west. **Achernar** (α Eridani) is skimming the southern horizon, but **Peacock** (α Pavonis) is now slightly higher and more easily visible. Although the **LMC** is roughly as high as the South Celestial Pole, the **SMC** and **47 Tucanae** are rather low (but still visible) in the south. In the east, the whole of **Scorpius** is now well clear of the horizon with the inconspicuous constellation of **Libra** preceding it along the ecliptic. The dense regions of the Milky Way in **Sagittarius** become visible later in the night.

Meteors

Two meteor showers, in particular, occur in April. The **π-Puppid** shower begins on April 15, but conditions are not particularly favourable in 2023 with shower maximum on April 23–24, when the Moon is 3–4 days old (a waxing crescent). The hourly rate is variable but the meteors tend to be faint. The parent body is the comet 26P/Grigg-Skjellerup. A second, more prolific, shower, the **η-Aquariids**, begins on April 19, a day before New Moon, and continues into May.

Omega Centauri (NGC 5139) is the largest and finest globular cluster in the sky (and in the Milky Way galaxy). It is believed to contain 10 million stars and differs in chemical composition and nature so greatly from other globulars that it may be the core of a disrupted dwarf galaxy, captured by the far more massive Galaxy.

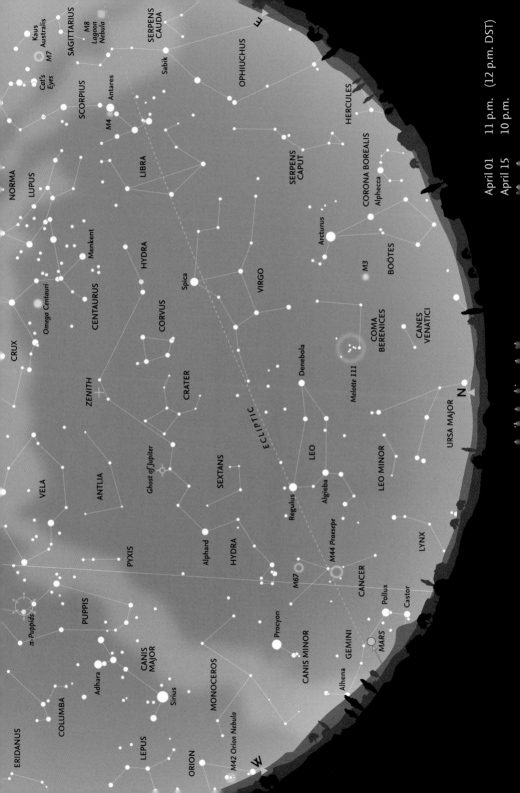

April 01   11 p.m.   (12 p.m. DST)
April 15   10 p.m.

April – Looking North

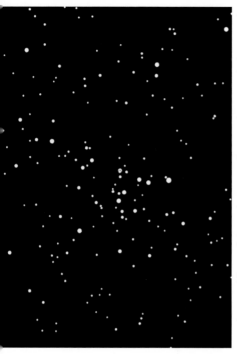

Leo is the most prominent constellation in the northern sky in April. *Gemini*, with *Castor* and *Pollux*, is low on the horizon in the west, and *Cancer* lies between the two constellations. To the east of Leo, the whole of *Virgo*, with *Spica* (α Virginis) its brightest star, is clearly visible, with the constellation of *Libra* farther east along the ecliptic. Above Leo and Virgo, the complete length of *Hydra* is visible, with *Alphard* (α Hydrae) forming a prominent triangle with *Regulus* and *Procyon* in *Canis Minor*. High in the sky, the small constellations of *Sextans* and *Crater*, together with the rather brighter *Corvus* lie between Leo, Virgo and Hydra.

Boötes and *Arcturus* are prominent in the northeastern sky, together with the globular cluster M3 in *Canes Venatici*. The circlet of *Corona Borealis* is close to the horizon. Between Leo and Boötes lies the constellation of *Coma Berenices*, notable for being the location of the open cluster Melotte 111 (this is sometimes called the Coma Cluster and confused with the Coma Cluster of galaxies (Abell 1656) east of *Denebola* in *Leo*). There are about 1,000 galaxies in Abell 1656, shown on the chart on page 61, and which is located near the North Galactic Pole, where we are looking out of the plane of the Galaxy and are thus

A very large, and frequently ignored, open star cluster, Melotte 111, also known as the Coma Cluster, is readily visible in the northern sky during April and May (south is up).

able to see deep into space. Only about ten of the brightest galaxies in the Coma Cluster are visible with the largest amateur telescopes.

The Moon's phases for April

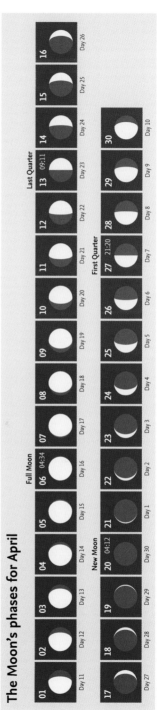

01	02	03	04	05	06 04:34	07
Day 11	Day 12	Day 13	Day 14	Day 15	Day 16	Day 17

Full Moon (above 06)

New Moon (above 20 04:12)

First Quarter (above 27 21:20)

Last Quarter (above 13 09:11)

April – Moon and Planets

The Moon

On April 2, the waxing gibbous Moon is 4.6° north of **Regulus** (mag. 1.4). On April 6, just after Full Moon, it is 3.3° north of **Spica** in **Virgo**. By April 10, it is 1.5° north of **Antares** in the morning sky. On April 16, it passes 3.5° south of **Saturn**, low in the evening sky. On April 19, one day before New Moon, it is 0.1° north of **Jupiter** in **Pisces**. On April 20, there is a hybrid **eclipse**, visible from Indonesia (see page 20). The next day, the Moon is 1.9° south of **Mercury** and, later, 1.7° north of **Uranus**, both too faint to be readily visible. By April 23 the waxing crescent Moon is 8.8° north of **Aldebaran** in **Taurus**. Later that day it is 1.3° north of brilliant **Venus** (mag. -4.1). On April 26, the Moon is 3.2° north of **Mars** in **Gemini** and then 1.5° south of **Pollux**. By April 29, it is again 4.6° north of **Regulus**, as it was at the beginning of the month.

The planets

Mercury is close to the Sun. It passed superior conjunction on March 17, and comes to greatest eastern elongation (19.5° from the Sun at mag. -0.0) on April 11. **Venus** is very bright (mag. -4.0 to -4.2) in the evening sky. **Mars** (mag. 1.1 to 1.3) moves across **Gemini** over the month. **Jupiter** is lost in the twilight in **Pisces**. **Saturn** (mag. 1.0.) is in **Aquarius**, very low in the morning twilight. **Uranus** (mag. 5.8 to 5.9) is slowly moving eastwards in **Aries**. **Neptune** is still in Pisces, at mag. 8.0 to 7.9.

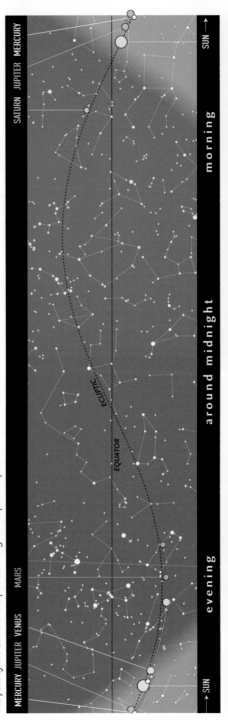

The path of the Sun and the planets along the ecliptic in April.

Calendar for April

Date	Time	Event
02		Australian and New Zealand Daylight Saving Time ends
02	08:01	Regulus 4.6°S of the Moon
06	04:34	Full Moon
06	17:23	Spica 3.3°S of the Moon
10	06:25	Antares 1.5°S of the Moon
11	22:10	Mercury at greatest elongation (19.5°E, mag. -0.0)
11	22:27	Jupiter at superior conjunction
13	09:11	Last Quarter
15–28		π-Puppid meteor shower
16	02:24	Moon at perigee = 367,968 km
16	03:49	Saturn 3.5°N of the Moon
17	17:24	Neptune 2.3°N of the Moon
19–May.28		η-Aquariid meteor shower
19	17:31	Jupiter 0.1°S of the Moon
20	04:12	New Moon
20	04:36	Hybrid solar eclipse
21	07:05	Mercury 1.9°N of the Moon
21	13:00	Uranus 1.7°S of the Moon
23	07:17	Aldebaran 8.8°S of the Moon
23	13:03	Venus 1.3°S of the Moon
24		π-Puppid meteor shower maximum
26	02:18	Mars 3.2°S of the Moon
26	18:03	Pollux 1.5°N of the Moon
27	21:20	First Quarter
28	06:43	Moon at apogee = 404,299 km
28	16:00	Regulus 4.6°S of the Moon

Evening 9 p.m.

April 2 • *The Moon is between Regulus and Algieba. Denebola (β Leo) is a little more than twenty degrees farther east.*

Evening 9 p.m.

April 5–6 • *In the eastern sky, the Moon moves; through the constellation of Virgo and passes Spica between April 5 and 6.*

Evening 10 p.m.

April 9–10 • *The Moon passes Antares, in the east-southeast.*

Evening 6 p.m.

April 23–24 • *The crescent Moon passes Venus and Elnath, with Aldebaran close by.*

Evening 6 p.m.

April 26 • *The Moon is between Mars and Castor; Pollux and Mars are equally bright (mag. 1.7.)*

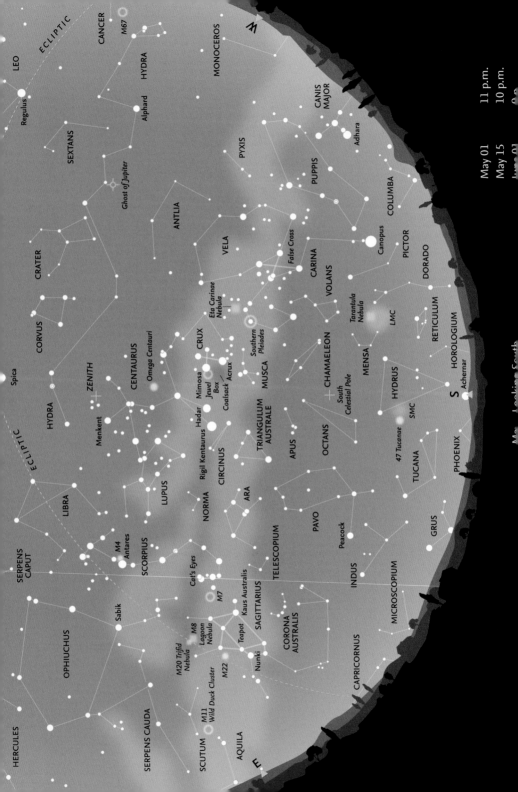

May — Looking South

May 01	11 p.m.
May 15	10 p.m.
June 01	9 p.m.

May – Looking South

In the west, **Canis Major** has set below the horizon, and **Canopus** (α Carinae) and **Puppis** are getting rather low. The whole of **Sagittarius** is now clearly seen in the east, with **Corona Australis** beneath it and **Scorpius** higher above. **Centaurus** is fully seen high in the south, with **Crux**, the **Eta Carinae Nebula**, the **Southern Pleiades** and the **False Cross** with **Vela** along the Milky Way to the west. In the south, **Achernar** (α Eridani) is skimming the southern horizon and the **Small Magellanic Cloud** (SMC) is beginning to rise higher in the sky, unlike the **Large Magellanic Cloud** (LMC) which is now lower. **Pavo** and **Peacock** (α Pavonis) are now much higher and even the faint constellations of **Tucana** and **Indus** are visible between Pavo and the southwestern horizon.

Meteors

The **η-Aquariids** are one of the two meteor showers associated with Comet 1P/Halley (the other being the **Orionids**, in October). The radiant is near the celestial equator, close to the 'Water Jar' in **Aquarius**, well below the horizon until late in the night (around dawn). However, meteors may still be seen in the eastern sky even when the radiant is below the horizon. There is a radiant map for the η-Aquariids on page 30. Their maximum in 2023, on May 6, occurs when the Moon is just after Full, so conditions are very unfavourable. Maximum hourly rate is about 50 per hour and a large proportion (about 25 per cent) of the meteors leave persistent trains.

Some faint constellations around the South Celestial Pole (SCP). Below the Pole, a part of the constellation of Octans is visible. Then, a little lower, Apus and farther down, Triangulum Australe. To the left is Musca and just above the centre of the image is Chamaeleon. The bright star near the top edge of the image is Miaplacidus, in the constellation of Carina.

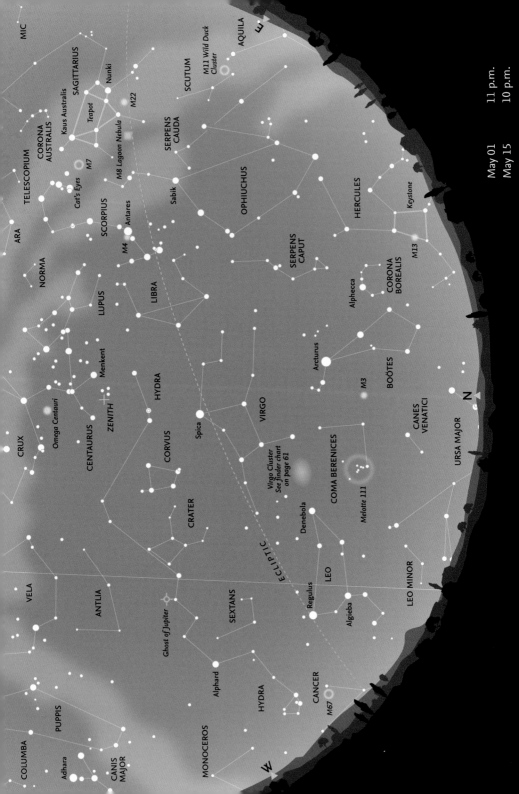

MIC
SAGITTARIUS
Kaus Australis
Nunki
Teapot
M22
CORONA AUSTRALIS
M8 Lagoon Nebula
M7
Cat's Eyes
TELESCOPIUM
SCORPIUS
Antares
M4
ARA
NORMA
LUPUS
CENTAURUS
Menkent
ZENITH
Omega Centauri
CRUX
VELA
PUPPIS
COLUMBA
Adhara
CANIS MAJOR
ANTLIA
Ghost of Jupiter
Alphard
HYDRA
MONOCEROS
SEXTANS
CANCER
M67
LEO
Regulus
Algieba
LEO MINOR
Denebola
COMA BERENICES
Melotte 111
CANES VENATICI
URSA MAJOR
BOÖTES
M3
Arcturus
CORONA BOREALIS
Alphecca
SERPENS CAPUT
HERCULES
Keystone
M13
OPHIUCHUS
Sabik
SERPENS CAUDA
SCUTUM
M11 Wild Duck Cluster
AQUILA
VIRGO
Spica
Virgo Cluster
See finder chart on page 61
CORVUS
CRATER
ECLIPTIC
LIBRA
HYDRA

N
E
S
W

May 01   11 p.m.
May 15   10 p.m.

May – Looking North

Boötes is almost due north, with brilliant, slightly orange-coloured **Arcturus** extemely prominent. The distinctive circlet of **Corona Borealis** is clearly visible to its east. The brightest star (α Coronae Borealis) is known as **Alphecca. Hercules** is rising in the east and becomes clearly visible later in the night. To the west of Boötes is the inconspicuous constellation of **Coma Berenices**, with the cluster **Melote 111** (see page 55) and, above it, the Virgo Cluster of galaxies (see here).

The large constellation of **Ophiuchus** (which actually crosses the ecliptic, and is thus the 'thirteenth' zodiacal constellation) is climbing into the eastern sky. Before the constellation boundaries were formally adopted by the International Astronomical Union in 1930, the southern region of Ophiuchus was regarded as forming part of the constellation of **Scorpius**, which had been part of the zodiac since antiquity.

Early in the night, the constellation of **Virgo**, with **Spica** (α Virginis), lies due north, with the rather faint constellation of **Libra** to its east. Farther along the ecliptic are Scorpius and brilliant, reddish **Antares** (α Scorpii). In the west, the constellation of **Leo** is readily visible now, and

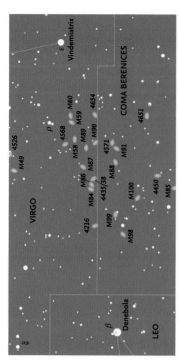

A finder chart for some of the brightest galaxies in the Virgo Cluster (see page 60). All stars brighter than magnitude 8.5 are shown (south is up).

both **Regulus** and **Denebola** (α and β Leonis, respectively) are prominent.

Virgo contains the nearest large cluster of galaxies, which is the centre of the Local Supercluster, of which the Milky Way galaxy forms part. The Virgo Cluster contains some 2,000 galaxies, the brightest of which are visible in amateur telescopes.

The Moon's phases for May

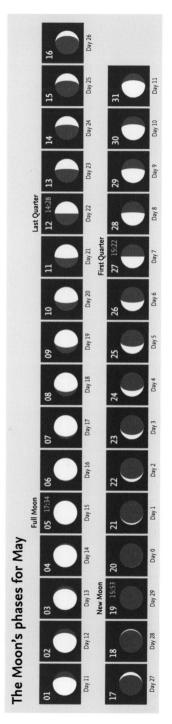

May – Moon and Planets

The Moon

On May 4, one day before Full Moon, the Moon is 3.3° north of *Spica* (mag. 0.98) in *Virgo*. On May 5 there is a penumbral lunar eclipse. Invisible to the naked eye, mid-eclipse is over the Indian Ocean. Two days later, the Moon is 1.5° north of *Antares*. By May 13, one day after Last Quarter, the Moon is 3.3° south of *Saturn* (mag. 1.0) in *Aquarius* in the morning sky. It passes 0.8° north of *Jupiter* in *Pisces* on May 17, with the planet lost in the morning twilight. On May 19, at New Moon, it is 1.8° north of *Uranus* (mag. 5.9) in *Aries*. The Moon is 8.7° north of *Aldebaran* on May 20 and 2.2° north of *Venus* on May 23. It passes 1.6° south of *Pollux* on May 24, and, later the same day, 3.8° north of *Mars.* The Moon is 4.6° north of *Regulus* in *Leo* on May 27. It is again 3.3° north of Spica on May 31.

The planets

Mercury is lost in twilight throughout May. *Venus* is bright (mag. -4.2 to -4.4), and initially in *Taurus*, then crosses into *Gemini*. There is an occultation of the planet on May 27, but this is essentially invisible, occurring in daylight and only visible from a small area of the southern Indian Ocean. *Mars* (mag. 1.3 to 1.6) moves from *Gemini* into *Cancer*, in the evening sky. *Jupiter* is inside *Pisces*, lost in the morning twilight. *Saturn* (mag. 1.0) is in *Aquarius*. *Uranus* is in *Aries* at mag. 5.9 and comes to superior conjunction on May 9. *Neptune* remains in *Pisces* at mag. 7.9.

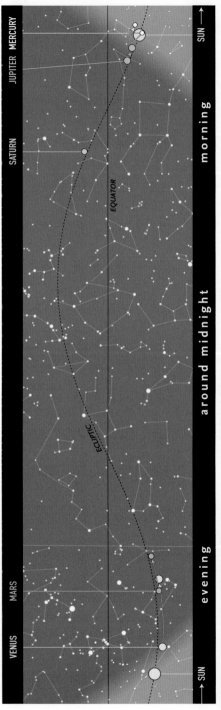

The path of the Sun and the planets along the ecliptic in May.

Calendar for May

01	23:28	Mercury at inferior conjunction
04	01:33	Spica 3.3° S of the Moon
05	17:22	Penumbral lunar eclipse
05	17:34	Full Moon
06		η-Aquariid meteor shower maximum
07	13:10	Antares 1.5° S of the Moon
09	19:56	Uranus at superior conjunction
11	05:05	Moon at perigee = 369,343 km
12	14:28	Last Quarter
13	13:07	Saturn 3.3°N of the Moon
15	01:25	Neptune 2.2°N of the Moon
17	13:18	Jupiter 0.8°S of the Moon
18	01:35	Mercury 3.6°N of the Moon
19	00:22	Uranus 1.8°S of the Moon
19	15:53	New Moon
20	15:55	Aldebaran 8.7°S of the Moon
23	12:08	Venus 2.2°S of the Moon
24	02:14	Pollux 1.6°N of the Moon
24	17:32	Mars 3.8°S of the Moon
26	01:39	Moon at apogee =404,509 km
27	00:09	Regulus 4.6°S of the Moon
27	15:22	First Quarter
29	05:34	Mercury at greatest elongation (24.9°W, mag. -0.5)
31	10:44	Spica 3.3°S of the Moon

Early morning 3 a.m.

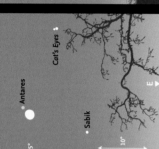
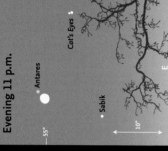

May 4–5 • *The Moon is almost Full when it passes north of Spica, in the constellation of Virgo.*

Evening 11 p.m.

May 7 • *High in the east, the Moon is close to Antares. Sabik and the Cat's Eyes are nearby.*

After midnight 2 a.m.

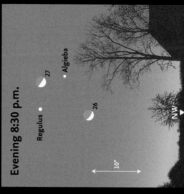

May 14 • *The Moon is near Saturn, in the east. Fomalhaut is farther south.*

Evening 6 p.m.

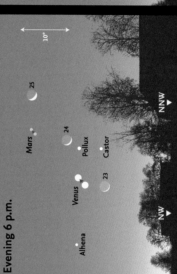
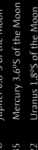
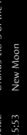
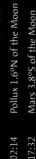

May 23–25 • *The crescent Moon passes Venus, Pollux and Castor and Mars. Alhena is farther west.*

Evening 8:30 p.m.

May 26–27 • *In the northwestern sky, the Moon passes between Regulus and Algieba.*

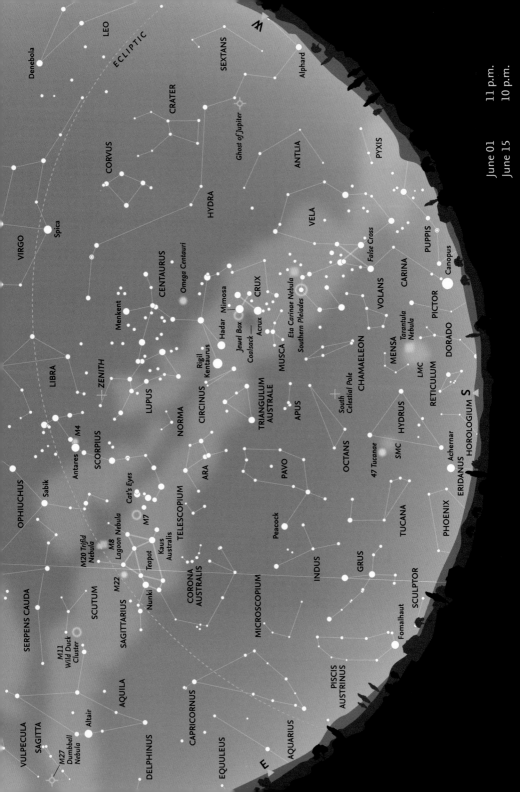

June – Looking South

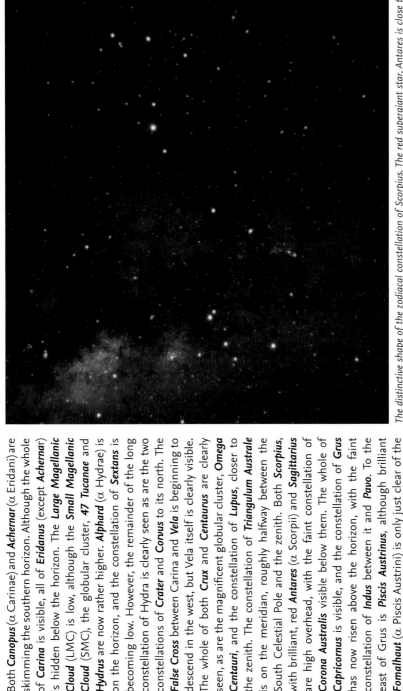

The distinctive shape of the zodiacal constellation of Scorpius. The red supergiant star, Antares is close to the zenith in June. The bright spot near the left side of the image is the open cluster M7, also known as Ptolomy's Cluster (north is up).

Both *Canopus* (α Carinae) and *Achernar* (α Eridani) are skimming the southern horizon. Although the whole of *Carina* is visible, all of *Eridanus* (except *Achernar*) is hidden below the horizon. The *Large Magellanic Cloud* (LMC) is low, although the *Small Magellanic Cloud* (SMC), the globular cluster, *47 Tucanae* and *Hydrus* are now rather higher. *Alphard* (α Hydrae) is on the horizon, and the constellation of *Sextans* is becoming low. However, the remainder of the long constellation of Hydra is clearly seen as are the two constellations of *Crater* and *Corvus* to its north. The *False Cross* between Carina and *Vela* is beginning to descend in the west, but Vela itself is clearly visible. The whole of both *Crux* and *Centaurus* are clearly seen, as are the magnificent globular cluster, *Omega Centauri*, and the constellation of *Lupus*, closer to the zenith. The constellation of *Triangulum Australe* is on the meridian, roughly halfway between the South Celestial Pole and the zenith. Both *Scorpius*, with brilliant, red *Antares* (α Scorpii) and *Sagittarius* are high overhead, with the faint constellation of *Corona Australis* visible below them. The whole of *Capricornus* is visible, and the constellation of *Grus* has now risen above the horizon, with the faint constellation of *Indus* between it and *Pavo*. To the east of Grus is *Piscis Austrinus*, although brilliant *Fomalhaut* (α Piscis Austrini) is only just clear of the horizon, and becomes clearly visible only later in the night and later in the month.

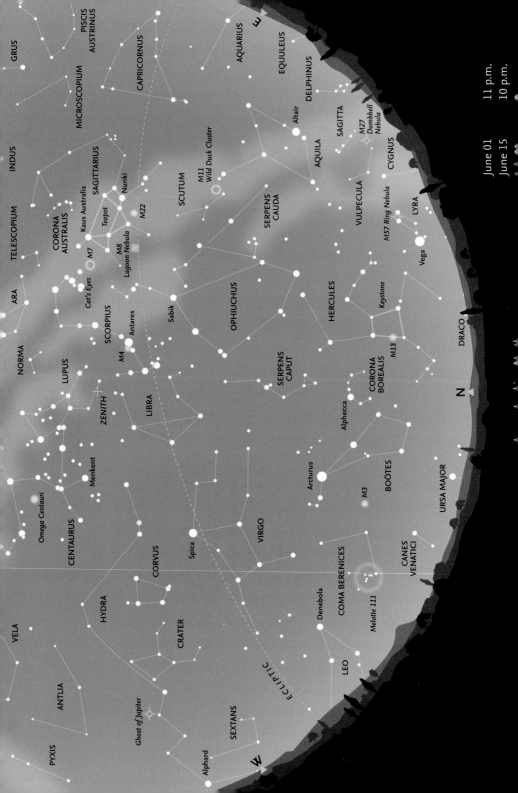

June – Looking North

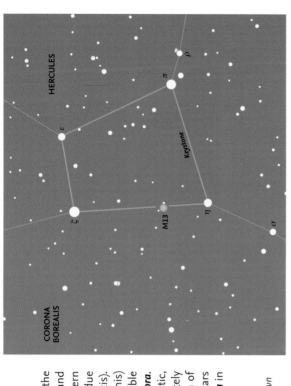

Vega in **Lyra** is just above the northeastern horizon. Closer to the meridian, the whole of **Hercules** is visible including the **'Keystone'** and **M13**, widely regarded as the finest globular cluster in the northern hemisphere. The small constellation of **Corona Borealis** is almost due north. To the west of the meridian is **Boötes** and **Arcturus** (α Boötis). Much of **Leo** is now below the western horizon, but **Denebola** (β Leonis) is still visible. Higher in the sky, the whole of **Virgo** is clearly visible and, still higher, not far from the zenith is the constellation of **Libra**. To its east is **Scorpius** and reddish **Antares**. Farther along the ecliptic, both the constellations of **Sagittarius** and **Capricornus** are completely visible. Between Hercules and Sagittarius is the large constellation of **Ophiuchus** and, to its east, **Aquila** and **Altair** (α Aquilae), one of the stars of the (northern) Summer Triangle. Another of the three stars, **Vega** in **Lyra**, is skimming the northern horizon.

Finder chart for M13, the finest globular cluster in the northern sky. All stars down to magnitude 7.5 are shown (south is up).

The Moon's phases for June

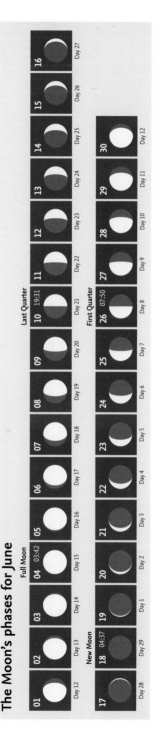

June – Moon and Planets

The Moon

On June 3, one day before Full Moon, the Moon is 1.5° north of *Antares* in *Scorpius*. By June 9, waning gibbous and one day before Last Quarter, it is 2.0° south of *Saturn* (mag. 0.9) in *Aquarius*. By June 11, it is 2.0° south of *Neptune* (mag. 7.9) in *Pisces*. On June 14, the Moon is 1.5° north of *Jupiter*, in *Aries*, low in the morning sky. The next day, June 15, it passes 2.0° north of *Uranus* (mag. 5.8) also in Aries. On June 20, the waxing crescent Moon is 1.7° south of *Pollux* in *Gemini*. By June 22, the Moon is 3.7° north of brilliant *Venus* (mag. -4.6) in *Cancer*. Later the same day it passes 3.8° north of *Mars*. The next day, June 23, the Moon is 4.4° north of *Regulus* in *Leo*. By June 27, it is 3.1° north of *Spica* in *Virgo*.

The planets

Mercury is low in the morning twilight. *Venus*, moving steadily eastwards, brightens from mag. -4.4 to -4.7 over the month. *Mars*, at mag. 1.6 to 1.7, is in *Cancer*, but moves into *Leo* later in the month. *Jupiter* is in *Aries* at mag. -2.1 to -2.2. *Saturn* (mag. 0.9 to 0.8) is moving slowly in *Aquarius*. *Uranus* remains in *Aries* at mag. 5.8 and *Neptune* is in *Pisces* at mag. 7.9.

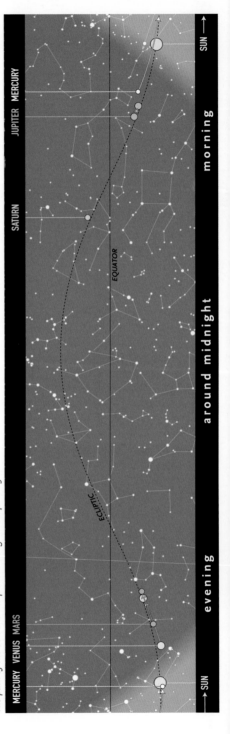

The path of the Sun and the planets along the ecliptic in June.

Calendar for June

03	21:53	Antares 1.5°S of the Moon
04	03:42	Full Moon
04	05:00 *	Uranus (mag. 5.8) 2.9°N of Mercury (mag. 0.1)
04	11:01	Venus at greatest elongation (45.4°E, mag. -4.4)
06	23:06	Moon at perigee = 364,861 km
09	20:23	Saturn 3.0°N of the Moon
10	19:31	Last Quarter
11	07:46	Neptune 2.0°N of the Moon
14	06:35	Jupiter 1.5°S of the Moon
15	09:54	Uranus 2.0°S of the Moon
16	20:38	Mercury 4.3°S of the Moon
16	23:09	Aldebaran 8.7°S of the Moon
18	04:37	New Moon
20	09:47	Pollux 1.7°N of the Moon
21	14:58	Winter solstice
22	00:48	Venus 3.7°S of the Moon
22	10:09	Mars 3.8°S of the Moon
22	18:30	Moon at apogee = 405,385 km
23	07:46	Regulus 4.4°S of the Moon
26	07:50	First Quarter
27	19:47	Spica 3.1°S of the Moon

* These objects are close together for an extended period around this time.

Morning 5 a.m.

June 4 • The Moon is almost Full when it passes Antares in the west-southwest. Sabik is almost due west.

Morning 5 a.m.

June 10 • In the north, at a latitude of 65°, Saturn is visited by the Moon. Fomalhaut is closer to the zenith.

Morning 6 a.m.

June 14–16 • In the early morning, the Moon passes Jupiter, Uranus (mag. 5.8) and the Pleiades, that will probably be lost in dawn.

Evening 5:30 p.m.

June 20–23 • The Moon passes Pollux (and Castor), Venus and Mars, and then passes between Regulus and Algieba. On June 20, both the Moon and Castor & Pollux will be lost in twilight.

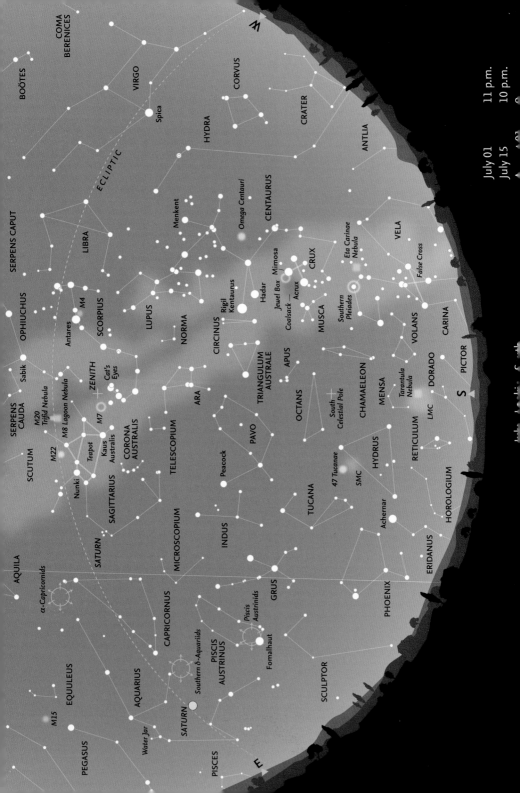

July – Looking South

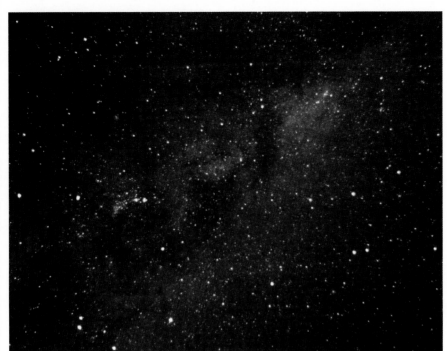

Part of the central Milky Way in the constellations of Norma and Scorpius, showing the dark clouds of obscuring dust (north is up).

Although the *Large Magellanic Cloud* (LMC) is almost due south, it is very low. Slightly farther west, the *False Cross* is nearing the horizon. Some of *Hydra* remains visible, but the constellation of *Crater* is becoming very low. *Corvus* is still easy to see. The zodiacal constellation of *Virgo* will soon be disappearing in the west. *Crux*, *Centaurus* and *Lupus* are still clearly seen, high in the sky, as is *Scorpius*, the tail of which is near the zenith. *Achernar* (α Eridani), *Hydrus* and the *Small Magellanic Cloud* (SMC) are now higher above the horizon and easier to observe. The whole of the constellation of *Phoenix* is also now clear of the horizon. Above it are *Tucana* and, halfway to the zenith, the constellation of *Pavo*. *Grus* and *Pisces Austrinus* (with brilliant *Fomalhaut*) are now fully visible. Higher in the sky are the zodiacal constellations of *Capricornus* and *Aquarius*, and even the westernmost portion of *Pisces* is rising above the horizon.

Meteors

July brings increasing meteor activity, mainly because there are several minor radiants active in the constellations of *Capricornus* and *Aquarius*. The first shower, the *α-Capricornids*, active from July 3 to August 15 (peaking July 30), does often produce very bright fireballs. The maximum rate, however, is only about 5 per hour. The parent body is probably that of the Comet 169P/NEAT. The most prominent shower is probably that of the *δ-Aquariids*, which are active from around July 12 to August 23, also with a peak on July 30, although even then the hourly rate is unlikely to reach 20 meteors per hour. In this case, the parent body is possibly Comet 96P/Machholz. This year, both shower maxima occur when the Moon is nearing Full, so observing conditions are particularly unfavourable. The *Piscis Austrinids* begin on July 15 and continue until August 10. Maximum is on July 28, but the rate is only about 5 per hour. The *Perseids* begin on July 17 and peak on August 12–13.

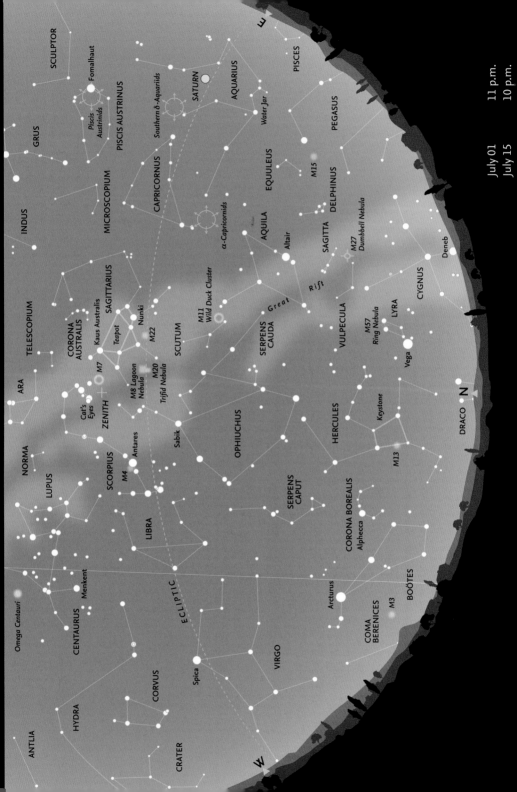

July – Looking North

The constellations of **Hercules** and **Lyra** are on either side of the meridian and both are clearly visible well above the horizon. To the east, **Deneb** (α Cygni) is skimming the horizon, but nearly all of the rest of the constellation is easily seen as is **Aquila** with **Altair** (α Aquilae). Between **Cygnus** and Aquila lies the small constellation of **Sagitta**, with **M27** (the Dumbbell Nebula, a planetary nebula). Even farther east, the whole extent of both the zodiacal constellations of **Aquarius** and **Capricornus** is visible. Above Aquila, farther along the **Great Rift** is the small constellation of **Scutum**, with **M11**, the Wild Duck Cluster. To its east is the tiny, but distinctive constellation of **Delphinus**. Still farther along the Great Rift lies the centre of the Milky Way Galaxy (in **Sagittarius**) and, near it, two emission nebulae: **M8** (the Lagoon Nebula) and **M20** (the Trifid Nebula). In the western sky, the constellation of **Boötes** and **Arcturus** (α Boötis) are beginning to approach the horizon, but **Corona Borealis** is still clearly seen. Even farther west, the whole of **Virgo** is visible, with **Libra** above it. The large constellation of **Ophiuchus** and the two halves of **Serpens** lie between Hercules and the zenith. **Scorpius** is draped around the actual zenith with Sagittarius to its east.

The Moon's phases for July

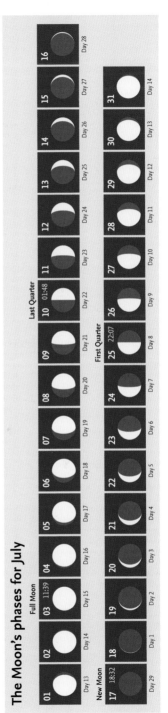

The constellation of Virgo. Spica (α Virginis) is the bright blue star near the top of the image, with Mars to its northeast (south is up).

July – Moon and Planets

The Earth

The Earth reaches aphelion, the farthest point from the Sun in its yearly orbit, at 20:07 on July 6 at a distance of 1.016680584 AU = 152,093,178.67 km.

The Moon

The Moon is 1.5° north of *Antares* in *Scorpius* on July 1, two days before Full Moon. On July 7, four days after Full, it is 2.7° south of *Saturn* (mag. 0.8) in *Aquarius*. By July 11, the Moon is 2.2° north of *Jupiter* (mag. -2.3) in *Aries*. On July 12 the waning crescent is 2.3° north of *Uranus* (mag. 5.8). By July 14 it is 8.8° north of *Aldebaran*. Three days later, at New Moon, it is 1.8° south of *Pollux* in *Gemini*. On July 19 it passes 3.5° north of *Mercury* (mag. -0.5), low in the evening twilight.

The next day the waxing crescent is 7.9° north of brilliant *Venus* (mag. -4.6), also low in the evening sky. Later that day, the Moon is 4.2° north of *Regulus*, between it and *Algieba* (γ Leonis). On July 21 the waxing crescent is 3.3° north of *Mars*, also in *Leo*. Before First Quarter on July 25 it is 2.8° north of *Spica* in *Virgo*. By July 28 it is 1.2° north of *Antares*.

The planets

Mercury, reaches superior conjunction on July 1, but is too close to the Sun to be readily visible. *Venus* is very bright (mag. -4.7 to -4.4) so may be visible although close to the Sun. *Mars* is mag. 1.7 to 1.8 in *Leo*. *Jupiter* is mag. -2.3 in *Aries*. *Saturn* is in *Aquarius*. *Uranus* at mag. 5.8 is in *Aries* and is on the border of *Taurus* at the end of the month. *Neptune* remains in *Pisces* at mag. 7.9.

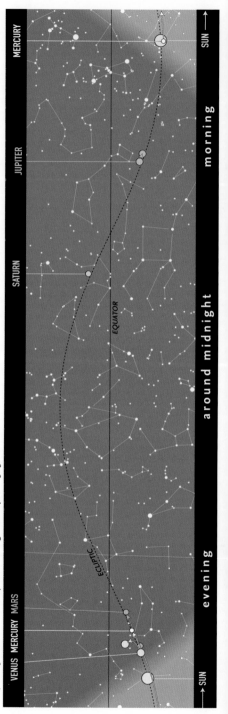

The path of the Sun and the planets along the ecliptic in July.

Calendar for July

01	05:06	Mercury at superior conjunction
01	07:55	Antares 1.5°S of the Moon
03–Aug.15		α–Capricornid meteor shower
03	11:39	Full Moon
04	22:25	Moon at perigee = 360,149 km
06	20:07	Earth at aphelion (1.01668 AU)
07	03:10	Saturn 2.7°N of the Moon
07	19:58	Minor planet (15) Eunomia at opposition (mag. 8.7)
08	14:12	Neptune 1.7°N of the Moon
10	01:48	Last Quarter
11	21:21	Jupiter 2.2°S of the Moon
12–Aug.23		Southern δ-Aquariid meteor shower
12	17:48	Uranus 2.3°S of the Moon
14	05:05	Aldebaran 8.8°S of the Moon
15–Aug.10		Piscis Austrinid meteor shower
17–Aug.24		Perseid meteor shower
17	16:22	Pollux 1.8°N of the Moon
17	18:32	New Moon
19	08:56	Mercury 3.5°S of the Moon
20	06:57	Moon at apogee = 406,289 km
20	08:37	Venus 7.9°S of the Moon
20	14:32	Regulus 4.2°S of the Moon
21	04:00	Mars 3.3°S of the Moon
25	03:41	Spica 2.8°S of the Moon
25	22:07	First Quarter
26	13:00 *	Mercury (mag. -0.2) 5.3°N of Venus (mag. -4.5)
28		Piscis Austrinid meteor shower maximum
28	17:46	Antares 1.2°S of the Moon
30		α–Capricornid meteor shower maximum
30		Southern δ-Aquariid meteor shower maximum

* These objects are close together for an extended period around this time.

Early morning 3 a.m.

July 1–2 • Near the west-southwest, the Moon passes Antares in the constellation of Scorpius. Sabik (η Oph) is farther west.

Early morning 3 a.m.

July 7–8 • In the north, about 25 degrees from the zenith, the Moon passes Saturn. Fomalhaut is even closer to the zenith.

Morning 5 a.m.

July 12–13 • In the northeast the Moon passes Jupiter, Uranus (mag. 5.8) and approaches the Pleiades.

Evening 5:45 p.m.

July 19–21 • After sunset the Moon passes Mercury, Venus, Regulus and Algieba. On July 21 it is near Mars.

Evening 5:45 p.m.

July 29 • Three planets join Regulus: Venus, Venus and Mars. The last one is also the faintest (mag. 1.8).

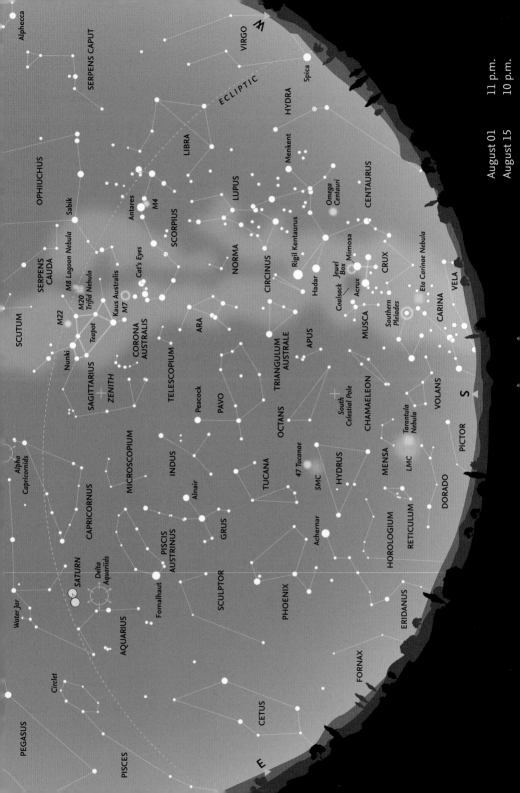

August – Looking South

Three inconspicuous constellations: *Volans*, *Chamaeleon* and *Octans* are on the meridian, with *Pavo* higher towards the zenith. Much of *Carina* is below the horizon, although the *Eta Carinae Nebula* and the *Southern Pleiades* are still visible. The *Large Magellanic Cloud* (LMC) and the faint constellation of *Mensa* are slightly higher in the sky. *Crux* is now lower, but the whole of *Centaurus* and *Lupus* remains visible. Much of *Virgo* has set in the west and *Libra* is following it down towards the horizon. *Scorpius* and *Sagittarius* are still visible high in the sky. *Achernar*, the *Small Magellanic Cloud* (SMC) and *Hydrus* are now well clear of the horizon, but below them are more small, inconspicuous constellations: *Dorado*, *Reticulum* and *Horlogium*. More of *Eridanus* is visible, together with parts of *Pictor* and *Fornax*. Much of *Cetus* has risen and the whole of the western arm of *Pisces* is now clearly seen. *Sculptor*, *Grus* and *Piscis Austrinus* lie halfway between the eastern horizon and the zenith, with faint *Microscopium* closer to the actual zenith.

Meteors

The *Piscis Austrinid* shower continues until about August 10, after maximum on July 28. The *Southern δ-Aquariids* reach maximum on July 30 to August 1, but are unlikely to exhibit more than about 20 meteors per hour. There are several minor southern showers active during the month: the *Northern δ-Aquariids* (maximum August 13), the *Southern* and *Northern ι-Aquariids* (maxima August 4 and August 19, respectively), but in all cases the rates are very low, just single figures per hour. The *α-Aurigid* shower begins on August 28, and is a short shower, lasting until September 5, reaching maximum in 2023 on September 1.

The Large Magellanic Cloud (LMC) lies mostly in the constellation of Dorado, partly extending into neighbouring Mensa. It contains many globular and open clusters as well as gaseous nebulae, such as the giant Tarantula Nebula.

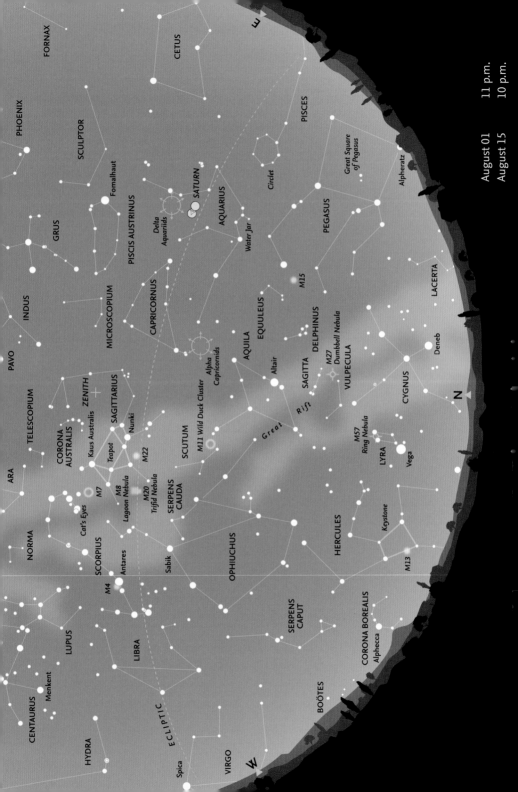

August – Looking North

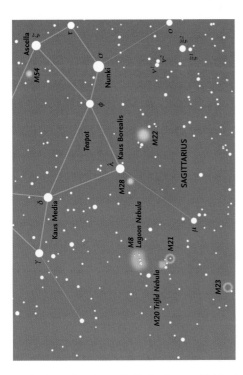

Cygnus, with brilliant Deneb (α Cygni), is now prominent just to the east of the meridian. The other two stars of the (northern) Summer Triangle, **Vega** in **Lyra** and **Altair** in **Aquila** are also unmistakeable in the sky. **Hercules**, however, is beginning to descend towards the northwestern horizon. Above it, the sprawling constellation of **Ophiuchus** is readily seen. The Great Square of **Pegasus** is now visible in the east, although **Alpheratz** (α Andromedae) is low on the horizon. The western side of **Pisces**, **Aquarius** and **Capricornus** are fully visible along the ecliptic. **Sagittarius** contains many gaseous nebulae, like **M8** (the Lagoon Nebula) and **M20** (the Trifid Nebula) and several globular clusters such as **M22**. The constellation is at the zenith, with Scorpius to the west.

A finder chart for the gaseous nebulae M8 (the Lagoon Nebula), M20 (the Trifid Nebula) and the globular cluster M22, all in Sagittarius. Clusters M21, M23 (open) and M28 (globular) are faint. The chart shows all stars brighter than magnitude 7.5 (south is up).

The Moon's phases for August

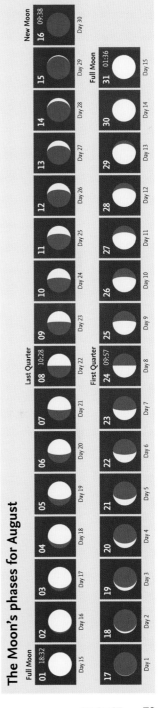

August – Moon and Planets

The Moon

On August 3, just after Full Moon, it is 2.5° south of **Saturn** in **Aquarius**. By August 8, waning gibbous, it is 2.9° north of **Jupiter** in **Pisces**. On August 9 it is 2.6° north of **Uranus** (mag. 5.8), close to **Taurus**. The next day, it is 9.0° north of **Aldebaran**. On August 13, a waning crescent, it is 1.7° south of **Pollux** in **Gemini**. On August 15, it is 13.3° north of **Venus** (mag. -4.1) in **Pisces** in the early evening sky. At New Moon on August 16, it is 4.1° north of **Regulus**. On August 18 it is 2.2° north of **Mars** in **Virgo**. By August 25, one day after First Quarter, it is 1.1° north of **Antares** in **Scorpius**. By August 30, one day before Full Moon, it is again 2.5° south of **Saturn** in **Aquarius**.

The Planets

Mercury is low in the twilight. It comes to greatest elongation (27.4°) east at mag. -0.3, on August 10. **Venus** (initially mag. -4.3, fades to mag. -4.1, then brightens to mag. -4.6) may be detectable in morning twilight. **Mars** (mag. 1.8) may be visible at the beginning of evening twilight. **Jupiter** is clearly visible in **Aries** at mag. -2.4 to -2.6. **Saturn**, in **Aquarius**, is mag. 0.4 at opposition on August 27. **Uranus** remains in **Aries** at mag. 5.9 and **Neptune** in **Pisces** at mag. 7.8.

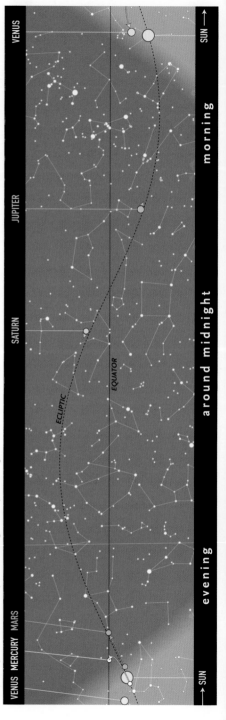

The path of the Sun and the planets along the ecliptic in August.

Morning 6 a.m.

10°

4

Saturn

3

Deneb Algedi

W ▶

August 3–4 • The Moon passes Saturn, in the western sky. On August 3, it is close to Deneb Algedi (δ Cap), a star of magnitude 2.9.

Early morning 4 a.m.

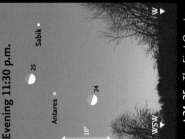

35°—

10°

8

Jupiter

Uranus
+

10

9

Pleiades

Hamal

Aldebaran

NNE ▶   NE ▶

August 8–10 • The Last Quarter Moon passes between Jupiter and Hamal. One day later it is close to The Pleiades, with Aldebaran farther east.

Morning 6:30 a.m.

10°

Castor

Pollux

Moon

Procyon

Alhena

NE ▶

August 14 • Just before sunrise you might be able to spot the narrow

Evening 11:30 p.m.

10°

25

24

Sabik

Antares

WSW ▶

August 24–25 • The First Quarter Moon passes Antares, with Sabik

Evening 10 p.m.

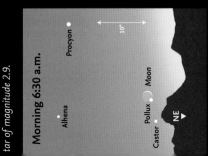

10°

• Deneb Algedi

30

31

Saturn

35°

E ▶

August 30–31 • High in the east, the Full Moon passes Deneb Algedi

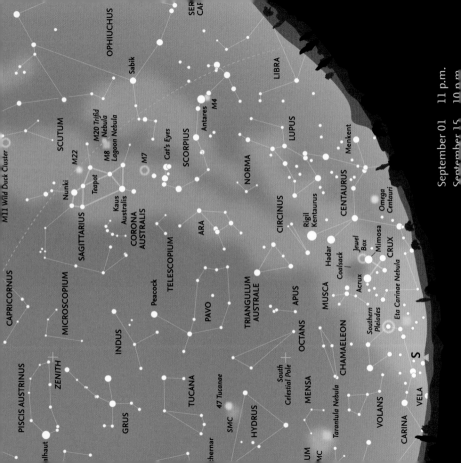

SER
CAP

OPHIUCHUS

Sabik

SCUTUM

M20 Trifid
Nebula
M8
Lagoon Nebula

M22

Cat's Eyes

M7

LIBRA

Antares

M4

SCORPIUS

M11 Wild Duck Cluster

Nunki

Teapot

Kaus
Australis

SAGITTARIUS

CORONA
AUSTRALIS

LUPUS

NORMA

Menkent

CENTAURUS

CIRCINUS

Omega
Centauri

Rigil
Kentaurus

CAPRICORNUS

MICROSCOPIUM

Peacock

TELESCOPIUM

ARA

Hadar

Jewel
Box

Mimosa

Coalsack

CRUX

INDUS

PAVO

TRIANGULUM
AUSTRALE

APUS

MUSCA

Acrux

Eta Carinae Nebula

PISCIS AUSTRINUS

ZENITH

OCTANS

Southern
Pleiades

GRUS

TUCANA

South
Celestial Pole

CHAMAELEON

S

alhaut

47 Tucanae

MENSA

VOLANS

VELA

SMC

HYDRUS

Tarantula Nebula

chernar

CARINA

UM

MC

September – Looking South

The smallest of the constellations, **Crux**, is now very low, but **Canopus** (α Carinae), the second brightest star in the sky, has now become visible once more. **Dorado** and the **Large Magellanic Cloud** (LMC) are higher and easier to see, as are the faint constellations of **Pictor, Caelum** and **Reticulum**. The **Small Magellanic Cloud** (SMC) and **47 Tucanae** are now nearly halfway between the horizon and the zenith. Although becoming low, the whole of **Centaurus** and **Lupus** remains visible in the southwest. **Scorpius** with reddish **Antares** is beginning to descend in the west, but **Sagittarius** is clearly seen high in the sky. The constellation of **Piscis Austrinus**, with **Fomalhaut** (α Piscis Austrini), is at the zenith. Below it are the constellations of **Grus** and **Pavo**, with **Achernar** (α Eridani) and almost the whole of the long constellation of **Eridanus.**

Meteors

There are no major meteor showers active in September. A few meteors may be seen from the **α-Aurigid** shower, active from late August, with possible maximum on September 1. The **Southern Taurid** shower begins this month (on September 10) and, although rates are low (less than 6 per hour), often produces very bright fireballs. This is a very long shower, lasting until about November 20. There is one minor shower, the **Piscids**, active throughout September with a slight maximum on September 19, but with an overall rate of no more than 3–5 meteors per hour. However, September is notable for a considerable increase in the number of sporadic meteors, which are, of course, completely unpredictable in location, direction and magnitude.

The 'Teapot' of Sagittarius at the top of the image with the curl of stars forming Corona Australis in the lower left, and the 'Cat's Eyes' (Shaula and Leseth) in the 'sting' of Scorpius near the right edge of the image. The bright open cluster M7 (in Scorpius) is prominent between Sagittarius and the Cat's Eyes (north is up).

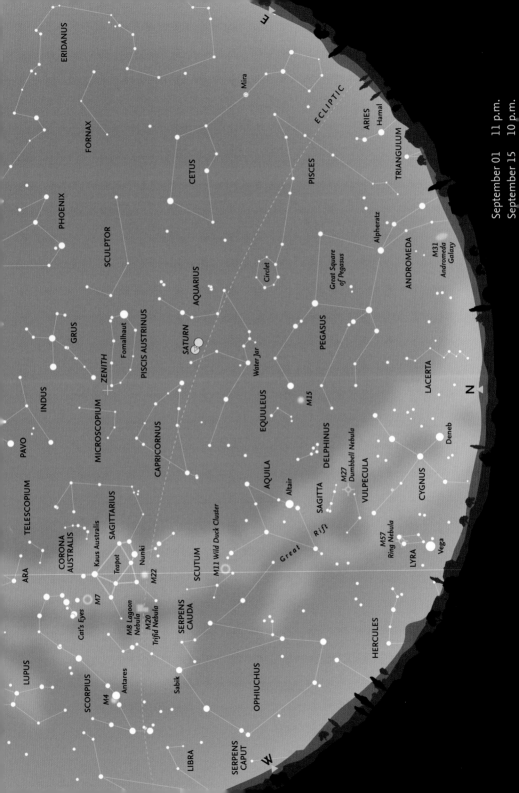

September 01   11 p.m.
September 15   10 p.m.

September – Looking North

The Great Square of *Pegasus* is nearing the meridian from the east and, on the west, the three bright stars of the Northern Summer Triangle, **Deneb** (α Cygni), **Altair** (α Aquilae) and **Vega** (α Lyrae) are still clearly seen, although Vega is low on the horizon, as is **M31** in **Andromeda**. **Delphinus** is well-placed, as are the fainter constellation of **Sagitta** and **Scutum** along the Milky Way. Giant **Ophiuchus** is beginning to descend towards the western horizon and much of **Libra** has disappeared. The three zodiacal constellations of *Pisces*, **Aquarius** and **Capricornus** are readily visible. Somewhat higher, **Piscis Austrinus** and **Fomalhaut** (α Piscis Austrini) is at the zenith. **Scorpius** and **Sagittarius** are now on the western side of the sky.

The constellations of Pisces and Aries. The 'Circlet' of Pisces is top left and the three main stars of Aries, bottom right. In 2023, Jupiter is in Aries from June until December (south is up).

The Moon's phases for September

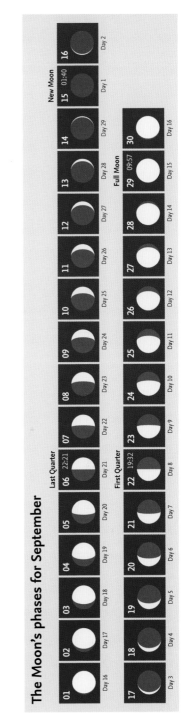

September – Moon and Planets

The Moon

On September 4, the waning gibbous Moon is 3.3° north of **Jupiter** (mag. -2.6) in **Aries**. The next day, it is 2.8° north of **Uranus** in **Aries**, which is much fainter at mag. 5.8. By September 6, at Last Quarter, it is 9.3° north of **Aldebaran** in **Taurus**. On September 10, the Moon is 1.6° south of **Pollux**. On September 11, it is 11.4° north of brilliant **Venus** at mag. -4.7 in **Cancer** in the morning sky. On September 13, it is 4.1° north of **Regulus** (mag. 1.4), which may just be visible in morning twilight. By September 16, one day after New Moon, it is 0.7° north of **Mars** in **Virgo**, at the end of evening twilight. The next day it is 2.4° north of **Spica**. By September 21, the Moon is 0.9° north of **Antares** in **Scorpius**. On September 27, the waxing gibbous Moon is 2.7° south of **Saturn** (mag. 0.5) in **Aquarius**.

The Planets

Mercury is generally lost in morning twilight, but comes to greatest western elongation (17.9°) at mag. -0.5 on September 22. **Venus** is in **Cancer**, visible in the morning sky at mag. -4.7. **Mars**, in **Virgo**, is lost in evening twilight. **Jupiter** begins retrograde motion in early September. It is in Aries at mag. -2.6 to -2.8. **Saturn** (mag. 0.5) is in **Aquarius**, and still retrograding. Uranus is just inside **Aries** at mag. 5.9 and **Neptune** is in **Pisces** at mag. 7.8 and comes to opposition on September 19.

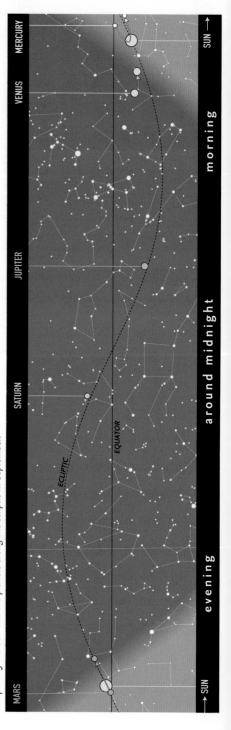

The path of the Sun and the planets along the ecliptic in September.

Calendar for September

1		α-Aurigid meteor shower maximum
1	07:21	Neptune 1.4°N of the Moon
4	19:47	Jupiter 3.3°S of the Moon
5	08:45	Uranus 2.8°S of the Moon
6	11:09	Mercury at inferior conjunction
6	17:21	Aldebaran 9.3°S of the Moon
6	22:21	Last Quarter
0–Nov.20		Southern Taurid meteor shower
0	04:10	Pollux 1.6°N of the Moon
1	12:59	Venus 11.4°S of the Moon
2	15:43	Moon at apogee = 406,291 km
3	02:39	Regulus 4.1°S of the Moon
3	17:40	Mercury 6.0°S of the Moon
5	01:40	New Moon
6	19:20	Mars 0.7°S of the Moon
7	15:51	Spica 2.4°S of the Moon
9	11:17	Neptune at opposition (mag. 7.8)
1	08:27	Antares 0.9°S of the Moon
2	13:16	Mercury at greatest elongation (17.9°W, mag. −0.5)
2	19:32	First Quarter
3	06:50	Spring equinox
4		New Zealand Daylight Saving Time begins
7	01:29	Saturn 2.7°N of the Moon
8	00:59	Moon at perigee = 359,911 km
8	16:59	Neptune 1.4°N of the Moon
9	09:57	Full Moon

Morning 5 a.m.

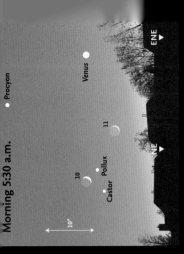

September 5–6 • In the northern sky, the Moon passes Jupiter, Uranus (mag. 5.7) and the Pleiades. Aldebaran and Hamal are also in the area.

Morning 5:30 a.m.

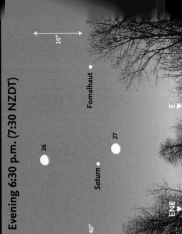

September 10–11 • The waning crescent Moon passes Pollux and Castor in the northeast. Venus is farther east.

Evening 9 p.m.

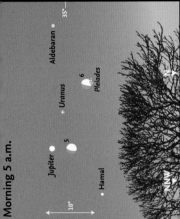

September 21 • The Moon is close to Antares, between it and Sabik. The Cat's Eyes may be found about fifteen degrees higher.

Evening 6:30 p.m. (7:30 NZDT)

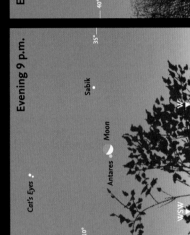

September 26–27 • High in the east, the Moon passes between Saturn and Fomalhaut. In New Zealand, Daylight Saving Time (NZDT) started on September 24.

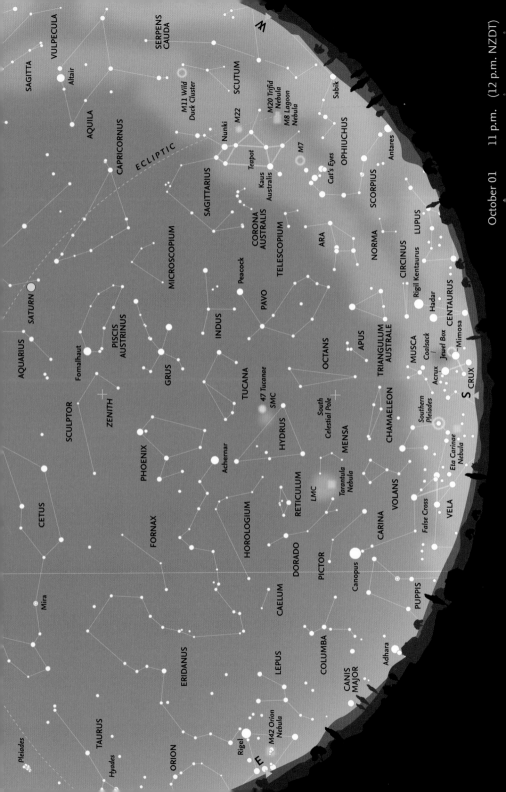

October 01   11 p.m.   (12 p.m. NZDT)

October – Looking South

Crux is now extremely low, only just visible above the horizon. The False Cross has reappeared, also very low, farther towards the east on the opposite side of the meridian. Canopus (α Carinae) is now much higher as is the **Large Magellanic Cloud** (LMC). The **Small Magellanic Cloud** (SMC) and the globular cluster **47 Tucanae** are on the southern meridian, halfway to the zenith as is **Achernar** (α Eridani). (The whole of the long constellation of **Eridanus** is now visible together with **Rigel** in **Orion**.) Still higher are **Phoenix, Grus** and **Piscis Austrinus. Pavo** has begun to descend in the southwest. **Ophiuchus** has now slipped below the horizon as has much of **Scorpius,** only the 'tail' of which remains visible. **Sagittarius** is getting lower, but remains visible, as do the zodiacal constellations of **Capricornus** and **Aquarius.**

Meteors

The **Orionids** are the major, fairly reliable meteor shower active in October. Like the May **η-Aquariid** shower, the Orionids are associated with Comet 1P/Halley. During this second pass through the stream of particles from the comet, slightly fewer meteors are seen than in May. In both showers the meteors are very fast, and many leave persistent trains. Although the Orionid maximum is nominally October 21–22, in fact there is a very broad maximum, lasting about a week from October 20 to 27, with hourly rates around 25. Occasionally, rates are higher (50–70 per hour). In 2023, the Moon is Full on October 28, so conditions at the end of maximum are very unfavourable.

The faint shower of the **Southern Taurids** (often with bright fireballs) peaks on October 10. The Southern Taurid maximum occurs when the Moon is a waning crescent, so conditions are reasonably favourable. Towards the end of the month (around October 20), another shower (the Northern Taurids) begins to show activity, which peaks early in November. The radiants for both Taurid showers are close together.

The combined position is shown on the 'Looking North' chart. The parent comet for both Taurid showers is Comet 2P/Encke.

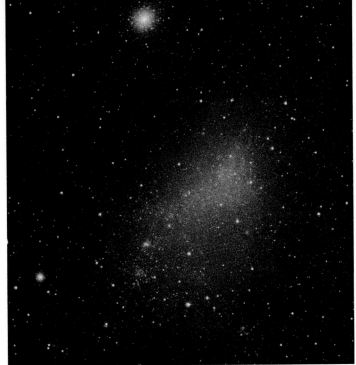

The Small Magellanic Cloud, one of the satellite galaxies of the Milky Way. Near the right edge of this image is 47 Tucanae (NGC 104), which is the second brightest globular cluster in the sky, after Omega Centauri (see page 53). North is up.

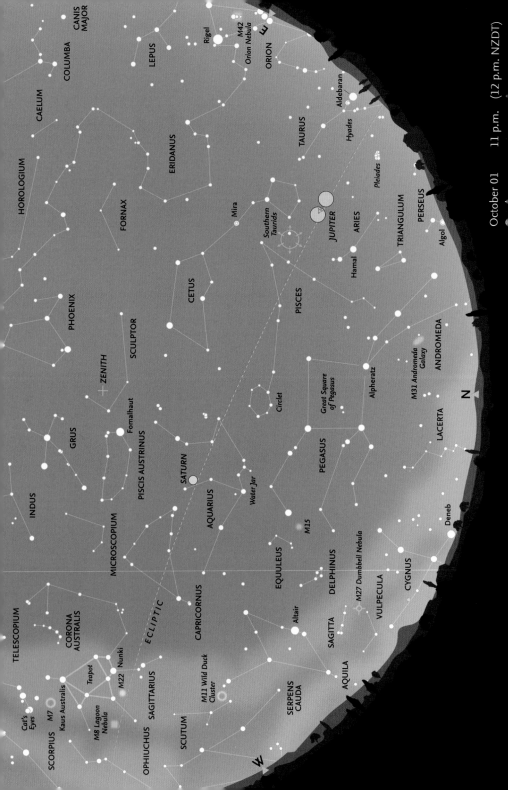

October 01   11 p.m.   (12 p.m. NZDT)

October – Looking North

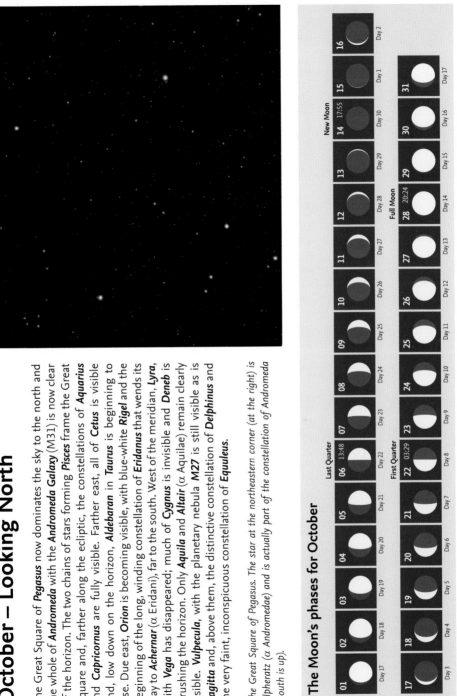

The Great Square of **Pegasus** now dominates the sky to the north and the whole of **Andromeda** with the **Andromeda Galaxy** (M31) is now clear of the horizon. The two chains of stars forming **Pisces** frame the Great Square and, farther along the ecliptic, the constellations of **Aquarius** and **Capricornus** are fully visible. Farther east, all of **Cetus** is visible and, low down on the horizon, **Aldebaran** in **Taurus** is beginning to rise. Due east, **Orion** is becoming visible, with blue-white **Rigel** and the beginning of the long, winding constellation of **Eridanus** that wends its way to **Achernar** (α Eridani), far to the south. West of the meridian, **Lyra**, with **Vega** has disappeared; much of **Cygnus** is invisible and **Deneb** is brushing the horizon. Only **Aquila** and **Altair** (α Aquilae) remain clearly visible. **Vulpecula**, with the planetary nebula **M27** is still visible as is **Sagitta** and, above them, the distinctive constellation of **Delphinus** and the very faint, inconspicuous constellation of **Equuleus.**

The Great Square of Pegasus. The star at the northeastern corner (at the right) is Alpheratz (α Andromedae) and is actually part of the constellation of Andromeda (south is up).

The Moon's phases for October

						Last Quarter					New Moon		
01	02	03	04	05	06 13:48	07	08	09	10	11	12	13	14 17:55	15	16
Day 17	Day 18	Day 19	Day 20	Day 21	Day 22	Day 23	Day 24	Day 25	Day 26	Day 27	Day 28	Day 29	Day 30	Day 1	Day 2

						First Quarter					Full Moon		
17	18	19	20	21	22 03:29	23	24	25	26	27	28 20:24	29	30	31
Day 3	Day 4	Day 5	Day 6	Day 7	Day 8	Day 9	Day 10	Day 11	Day 12	Day 13	Day 14	Day 15	Day 16	Day 17

October – Moon and Planets

The Moon

On October 2, the Moon passes 3.4° north of *Jupiter* (mag. -2.8), retrograding in *Aries*. Later that day it is 2.9° north of *Uranus* (mag. 5.7). It then passes the *Pleiades* and on October 4 is 9.4° north of *Aldebaran*. By October 7, one day after Last Quarter, it is 1.4° south of *Pollux* in *Gemini*. On October 10, the waning crescent is 4.2° north of *Regulus* in *Leo*. Later the same day it is 6.5° north of brilliant *Venus* (mag. -4.6) also in Leo. On October 14, at New Moon, there is an annular solar eclipse (see pages 20–21). Later that day, the Moon is 2.4° north of *Spica* in twilight. By October 18, it is 0.8° north of *Antares* in *Scorpius*. On October 24, the Moon is 2.8° south of *Saturn* in *Aquarius*. By October 29, one day after Full, the Moon is 3.1° north of *Jupiter* and on October 30

it is 2.9° north of *Uranus*, both in *Aries*. The next day, October 31, the waning gibbous Moon passes 9.4° north of *Aldebaran*.

The planets

Mercury is lost in twilight close to the Sun, and is at superior conjunction on October 20. *Venus*, in *Leo*, is bright (mag. -4.7 to -4.4) in the evening sky and moving closer to the Sun. *Jupiter* is retrograding in *Aries* at mag. -2.8 to -2.9. *Saturn* is also retrograding in *Aquarius*. *Uranus* is still in *Aries* at mag. 5.8 and *Neptune* is slowly retrograding in *Pisces* at mag. 7.8.

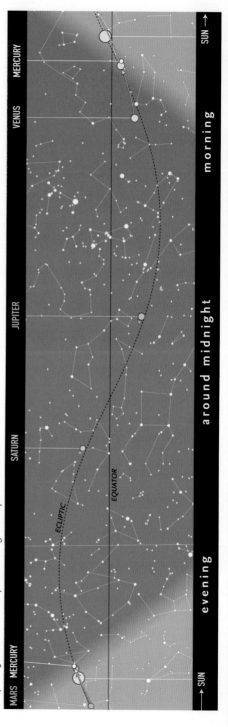

The path of the Sun and the planets along the ecliptic in October.

Calendar for October

Morning 4 a.m. (DST)

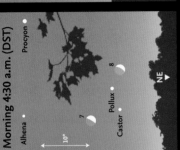

Jupiter · Uranus · Pleiades · Hamal · 2 · 3 · NNW · 10° · 35°

October 2–3 • *The Moon passes below Jupiter and Uranus (mag. 5.7) and approaches the Pleiades.*

Evening 10 p.m. (DST)

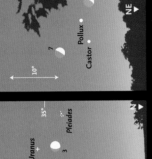

ZENITH · Fomalhaut · Moon · Saturn · N · 10°

October 24 • *Very high in the north, the Moon is between Saturn and Fomalhaut*

Morning 4:30 a.m. (DST)

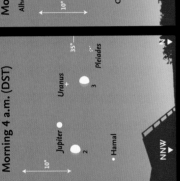

Alhena · Procyon · Castor · Pollux · 7 · 8 · NE · 10°

October 7–8 • *The Moon passes Castor and Pollux in the northeast. Alhena and Procyon are near.*

Evening 11 p.m. (DST)

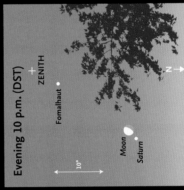

Jupiter · Hamal · Uranus · Pleiades · Aldebaran · 28 · 29 · 30 · NNE · NE · 35° · 10°

October 28–30 • *The Moon passes Hamal, Jupiter, Uranus (mag. 5–6) and the Pleiades. Aldebaran is farther east*

Morning 5:45 a.m. (DST)

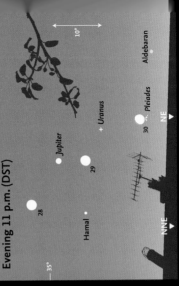

Venus · Regulus · Algieba · 10 · 11 · ENE · 10°

October 10–11 • *In the morning twilight, the crescent Moon passes below Venus and Regulus.*

1		Australian Daylight Saving Time begins
2–Nov.07		Orionid meteor shower
02	03:20	Jupiter 3.4°S of the Moon
02	17:15	Uranus 2.9°S of the Moon
04	01:46	Aldebaran 9.4°S of the Moon
06	13:48	Last Quarter
07	11:02	Pollux 1.4°N of the Moon
0		Southern Taurid meteor shower maximum
0	03:42	Moon at apogee = 405,426 km
0	09:20	Regulus 4.2°S of the Moon
0	09:44	Venus 6.5°S of the Moon
4	09:33	Mercury 0.7°N of the Moon
4	17:55	New Moon
4	17:59	Annular solar eclipse
4	22:07	Spica 2.4°S of the Moon
5	16:17	Mars 1.0°N of the Moon
8	13:53	Antares 0.8°S of the Moon
0–Dec.10		Northern Taurid meteor shower
0	05:38	Mercury at superior conjunction
1		Orionid meteor shower maximum
2	03:29	First Quarter
3	23:14	Venus at greatest elongation (46.4°W, mag. -4.5)
4	07:56	Saturn 2.8°N of the Moon
6	01:23	Neptune 1.5°N of the Moon
6	03:02	Moon at perigee = 364,872 km
8	20:13	Partial lunar eclipse
8	20:24	Full Moon
29	01:00	European Daylight Saving Time (British Summer Time, BST) ends
9	08:14	Jupiter 3.1°S of the Moon
9	17:00 ★	Mars (mag. 1.5) 0.4°N of Mercury (mag. -1.0)
0	01:53	Uranus 2.9°S of the Moon
1	11:28	Aldebaran 9.4°S of the Moon

se objects are close together for an extended period

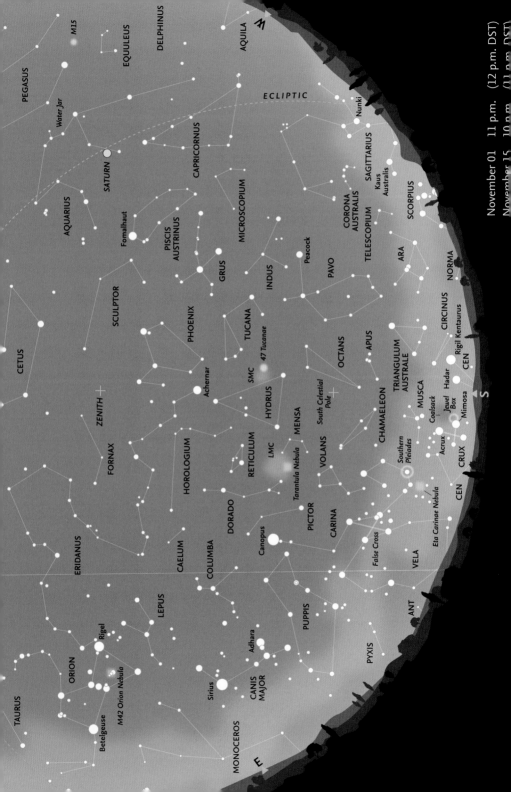

November 01 11 p.m. (12 p.m. DST)
November 15 10 p.m. (11 p.m. DST)

November – Looking South

Crux and the two brightest stars of **Centaurus**, **Rigil Kentaurus** (α Centauri) and **Hadar** (β Centauri), are extremely low on the southern horizon. The **False Cross** on the **Carina/Vela** border is now higher, and the constellation of **Puppis** as well as **Canopus** (α Carinae) and both the **Large Magellanic Cloud** (LMC) and the **Small Magellanic Cloud** (SMC) are clearly visible. **Achernar** (α Eridani) is halfway between the South Celestial Pole and the zenith. The whole of **Eridanus**, which starts near **Rigel** in **Orion**, is now clearly seen as it winds its way to Achernar. **Pavo** is becoming lower in the southwest, and in the west, most of **Sagittarius** is below the horizon, with **Capricornus** descending behind it. **Corona Australis** is still just visible. In the east, **Canis Major** is now clearly seen, together with the small constellations of **Columba** and **Lepus** above it.

Meteors

Two meteor showers begin in September or October but continue into November. The **Orionids** (see page 89), one of the streams associated with Comet 1/P Halley, continue until about November 7. The **Southern Taurids** (see page 83) begin on September 10 and continue until November 20. Because of the location of the radiant, the **Leonid** shower is best seen from the northern hemisphere, but southern observers may see some rising from the horizon. They have a short period of activity (November 6–30), with maximum on November 17–18. This shower is associated with Comet 55P/Tempel-Tuttle and has shown extraordinary activity on various occasions with many thousands of meteors per hour. The rate in 2023 is likely to be about 10 per hour. These meteors are the fastest shower meteors recorded (about 70 km per second) and often leave persistent trains. The shower is very rich in faint meteors.

There is a minor southern meteor shower that begins activity in late November (nominally November 28). This is the **Phoenicids**, but little is

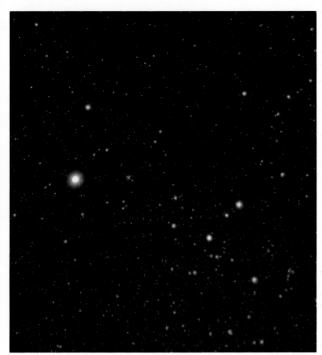

The constellation of Canis Major, with Sirius, the brightest star in the sky (mag. -1.4) in the upper half of the image. Below Sirius is the open cluster M41 (north is up).

known of the shower, partly because the parent comet is believed to be the disintegrated comet D/1819 W1 (Blanpain). With no accurate knowledge of the location of the remnants of the comet, predicting the possible rate becomes little more than guesswork, but the rate is variable and may rapidly increase (as might be expected) if the orbit is nearby. Bright meteors tend to be quite frequent and the meteors are fairly slow. The radiant is located within Phoenix, not far from the border with **Eridanus** and the bright star **Achernar** (α Eridani).

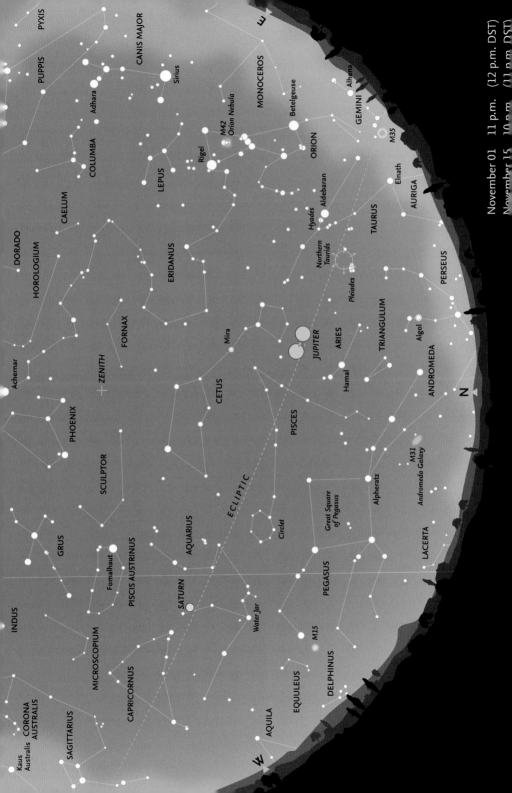

November – Looking North

Andromeda is now due north, and the **Andromeda Galaxy** (M31) has risen sufficiently to be clearly seen. The constellation of **Triangulum** lies between Andromeda and the zodiacal constellation of **Aries**. Much of **Perseus** (including the variable star, **Algol**) is now above the horizon, together with part of **Auriga**. The whole of **Pegasus**, with the **Great Square**, is clearly seen and, above it the two lines of stars forming **Pisces** and the distinctive asterism of the **Circlet**. Higher still is the constellation of **Cetus** with the famous variable star, **Mira**. The whole of **Taurus** with **Aldebaran** (α Tauri), and the **Pleiades** and the **Hyades** clusters are visible in the southeast. **Orion** has fully risen in the east with **Lepus** above it. The long, winding constellation of **Eridanus** begins near **Rigel** in Orion. The faint constellation of **Monoceros** lies to the east of Orion and straddles the Milky Way. **Canis Major** and brilliant **Sirius** are even farther round towards the east. In the west, the zodiacal constellations of **Aquarius** (with its distinctive asterism of the **Water Jar**) and **Capricornus** are clearly seen, with **Piscis Austrinus** and bright **Fomalhaut** (α Piscis Austrini) higher in the sky. The faint constellation of **Sculptor** lies between Piscis Austrinus and the zenith.

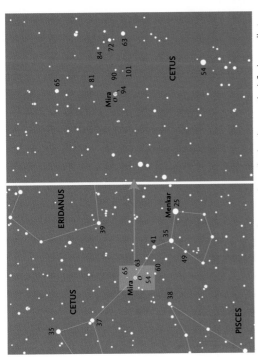

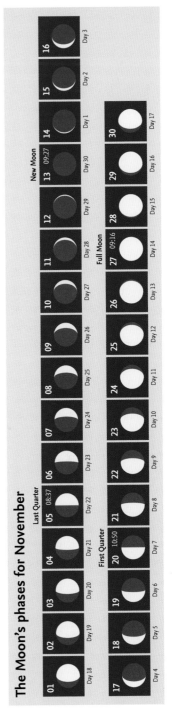

Finder and comparison charts for Mira (o Ceti). The chart on the left shows all stars brighter than magnitude 6.5. The chart on the right shows stars down to magnitude 10.0. The comparison star magnitudes are shown without the decimal point (south is up).

The Moon's phases for November

November – Moon and Planets

The Moon

On November 3, the waning gibbous Moon is 1.5° south of **Pollux**. On November 6 it passes 4.2° north of **Regulus**, between it and **Algieba**. By November 9 it is 1.0° north of **Venus** (mag. -4.4) in **Virgo**. On November 11 it passes 2.4° north of **Spica** in **Virgo**. By November 20, the Moon is 2.7° south of **Saturn** in **Aquarius**. On November 25 it passes 2.8° north of **Jupiter** in **Aries**, then the next day, 2.6° north of **Uranus** (mag. 5.6) also in **Aries**. On November 27 it is 9.3° north of **Aldebaran**.

The Planets

Mercury is in evening twilight, but may become visible at the end of the month on the border of **Ophiuchus** and **Sagittarius** at mag. -0.4 to -0.5. **Venus** (mag. -4.4 to -4.2) begins in **Leo** and moves across **Virgo**. **Mars** is lost in twilight near the Sun. **Jupiter** (mag. -2.9) comes to opposition in **Aries** on November 3. **Saturn** (mag. 0.7 to 0.8) initially retrograding, resumes direct motion on November 5. **Uranus** (mag. 5.6) is at opposition in **Aries** on November 13 and **Neptune** is slowly retrograding in **Pisces** at mag. 7.9.

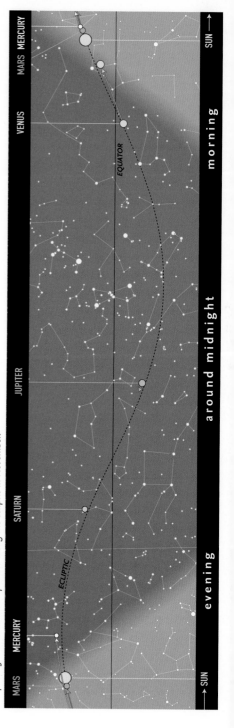

The path of the Sun and the planets along the ecliptic in November.

Calendar for November

03	05:02	Jupiter at opposition (mag. -2.9)
03	19:09	Pollux 1.5°N of the Moon
05	08:37	Last Quarter
06–30		Leonid meteor shower
06	16:59	Regulus 4.2°S of the Moon
06	21:49	Moon at apogee = 404,569 km
09	09:30	Venus 1.0°S of the Moon
11	05:48	Spica 2.4°S of the Moon
12–13		Northern Taurid meteor shower maximum
13	09:27	New Moon
13	13:32	Mars 2.5°N of the Moon
13	17:21	Uranus at opposition (mag. 5.6)
14	14:39	Mercury 1.7°N of the Moon
14	20:18	Antares 0.9°S of the Moon
17–18		Leonid meteor shower maximum
18	05:41	Mars at superior conjunction
20	10:50	First Quarter
20	14:06	Saturn 2.7°N of the Moon
21	21:01	Moon at perigee = 369,818 km
22	07:45	Neptune 1.5°N of the Moon
25	11:14	Jupiter 2.8°S of the Moon
26	09:19	Uranus 2.6°S of the Moon
27	09:16	Full Moon
27	21:03	Aldebaran 9.3°S of the Moon
28–Dec.09		Phoenicid meteor shower

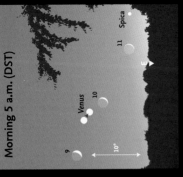

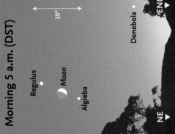

Morning 5 a.m. (DST)

November 4 • Shortly before Last Quarter, the Moon is close to Pollux and Castor.

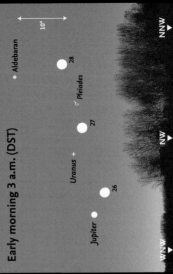

Morning 5 a.m. (DST)

November 7 • The Moon is between Regulus and Algieba. Denebola is close to the horizon.

Morning 5 a.m. (DST)

November 9–11 • The Moon passes Venus and Spica, low in the east. Spica is probably lost in twilight.

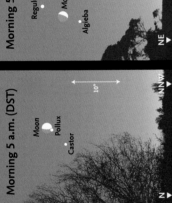

Midnight (DST)

November 20/21 • At midnight the First Quarter Moon and Saturn are side by side in the west. Deneb Algedi is also near.

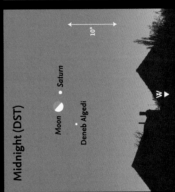

Early morning 3 a.m. (DST)

November 26–28 • In the northwest, the Moon passes Jupiter, Uranus (mag. 5.6) and the Pleiades. Aldebaran is nearby.

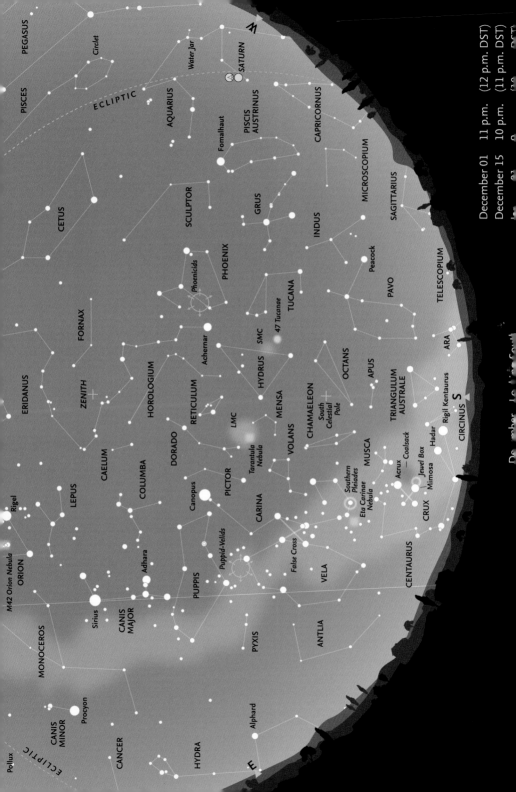

PEGASUS

Circlet

PISCES

ECLIPTIC

Water Jar

SATURN

AQUARIUS

PISCIS
AUSTRINUS

CETUS

Fomalhaut

SCULPTOR

CAPRICORNUS

GRUS

MICROSCOPIUM

Phoenicids

PHOENIX

INDUS

SAGITTARIUS

ERIDANUS

FORNAX

Achernar

SMC

TUCANA

47 Tucanae

Peacock

PAVO

TELESCOPIUM

ZENITH

HOROLOGIUM

RETICULUM

HYDRUS

OCTANS

ARA

CAELUM

DORADO

LMC

MENSA

South
Celestial
Pole

APUS

S

CIRCINUS

CHAMAELEON

TRIANGULUM
AUSTRALE

LEPUS

COLUMBA

PICTOR

Tarantula
Nebula

VOLANS

Rigil Kentaurus

Hadar

Rigel

Canopus

CARINA

MUSCA

Coalsack

M42 Orion Nebula

Southern
Pleiades

Acrux

Jewel Box

ORION

Eta Carinae
Nebula

Mimosa

CRUX

Adhara

Puppid-Velids

CENTAURUS

MONOCEROS

Sirius

CANIS
MAJOR

PUPPIS

False Cross

VELA

ANTLIA

Procyon

PYXIS

CANIS
MINOR

CANCER

Alphard

Pollux

HYDRA

ECLIPTIC

E

December 01   11 p.m.   (12 p.m. DST)
December 15   10 p.m.   (11 p.m. DST)

December – Looking South

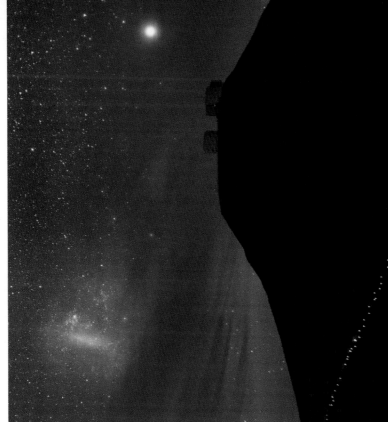

The Large Magellanic Cloud seen over the European Southern Observatory at Paranal in Chile. Brilliant Canopus (α Carinae), the second brightest star in the southern hemisphere, is visible through the low cloud to the right.

Crux and the two brightest stars in **Centaurus**, **Hadar** and **Rigil Kentaurus**, are now higher above the horizon. The **Eta Carina Nebula** and the **Southern Pleiades** are now conveniently placed for observation. Above them, the **False Cross** is clearly seen, with the whole of **Vela** and below it the inconspicuous constelation of **Antlia**. **Carina** with **Canopus** (α Carinae) and **Puppis** are roughly halfway between the horizon and the zenith. **Sirius** and **Canis Major** are high in the east. In the west, **Achernar** (α Eridani) and **Phoenix** are about the same altitude as Canopus. The faint constellations of **Pictor**, **Dorado**, **Reticulum** and **Horlogium** lie between them. **Pavo** with **Peacock** (α Pavonis) is becoming low, as are the constellations of **Indus**, **Grus** and **Piscis Austrinus**. Higher still are the inconspicuous constellations of **Sculptor** and, near the zenith, **Fornax**. **Capricornus** is largely invisible, but most of **Aquarius** may still be seen.

Meteors

The **Phoenicid** shower continues into December, reaching its weak maximum on December 2. The **Puppid Velid** shower's radiant is on the border between the two constellations. The shower begins on December 1, lasting until December 15, with maximum on December 7. It is a weak shower with a maximum hourly rate of about 10 meteors, but bright meteors are often seen. One of the most dependable showers of the year is the **Geminids**, active from December 4 to 20, with maximum in 2023 on December 14–15. The rate is often 60–70 per hour, and may rise even higher.

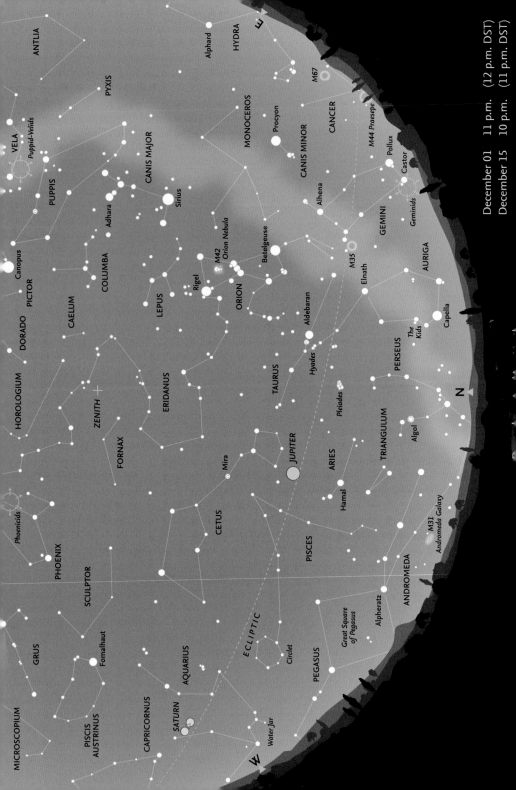

December 01   11 p.m.   (12 p.m. DST)
December 15   10 p.m.   (11 p.m. DST)

December – Looking North

Most of **Perseus** may be seen due north with, above it, the **Pleiades** cluster. Farther west, **Andromeda** is very low and the southern stars of **Pegasus** have been lost below the horizon. Above Pegasus lies the zodiacal constellation of **Pisces** and, still higher, **Cetus** and the faint constellatons of **Sculptor** and **Fornax**. The famous variable of **Mira** in Cetus (see charts on page 97) is ideally placed for observation. Towards the east, the whole of the constellations of both **Auriga** and **Gemini** is visible, although **Capella** (α Aurigae) and **Castor** and **Pollux** (α and β Gemini) are low on the horizon. **Taurus, Orion** and **Canis Minor** are readily visible, together with the faint constellation of **Monoceros**. **Eridanus** wanders from its start near **Rigel** (β Orionis) towards **Achernar** (α Eridani), beyond the zenith.

The constellation of Taurus contains two contrasting open clusters: the compact Pleiades, with its striking blue-white stars, and the more scattered, 'V'-shaped Hyades, which are much closer to us. Orange Aldebaran (α Tauri) is not related to the Hyades, but lies between it and the Earth (south is up).

The Moon's phases for December

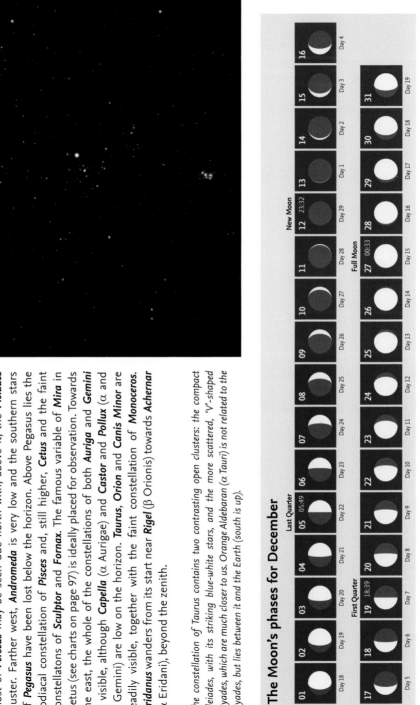

December – Moon and Planets

The Moon

On December 1, the Moon is 1.6° south of *Pollux* in *Gemini*. On December 4 the Moon passes 4.0° north of *Regulus* between it and *Algieba*. By December 8 the waning crescent Moon is 2.3° north of *Spica*. The next day it is 3.6° south of *Venus*. By New Moon on December 12, lost in twilight, it is 0.9° north of *Antares* and later that day, 3.6° south of *Mars*. On December 17 the Moon is 2.5° south of *Saturn* in *Aquarius*. By December 22 the waxing gibbous Moon is 2.6° north of *Jupiter* in *Aries*. It is 2.8° north of Uranus (mag. 5.7), also in Aries, the next day. It passes 9.4° north of *Aldebaran* on December 25. One day after Full Moon, on December 28, it is 1.7° south of *Pollux* and on December 31 3.8° north of *Regulus*, again between it and *Algieba*.

The planets

Mercury is at greatest eastern elongation (21.3°) on December 4 at mag -0.5. *Venus* (mag. -4.1) moves from *Virgo* into *Libra* and morning twilight. *Mars* remains too close to the Sun to be visible. *Jupiter* (mag. -2.8 to -2.6) is in *Aries*. *Saturn* is moving slowly eastwards in *Aquarius*. *Uranus* is in *Aries* and *Neptune* is still slowly retrograding in Pisces at mag. 7.9. The minor planet **(4)** *Vesta* is at opposition (mag. 6.4) on December 21 (see page 26) and **(9)** *Metis* (mag. 8.4) at opposition the next day in *Gemini*.

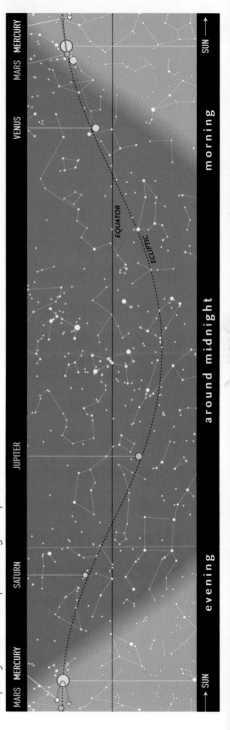

The path of the Sun and the planets along the ecliptic in December.

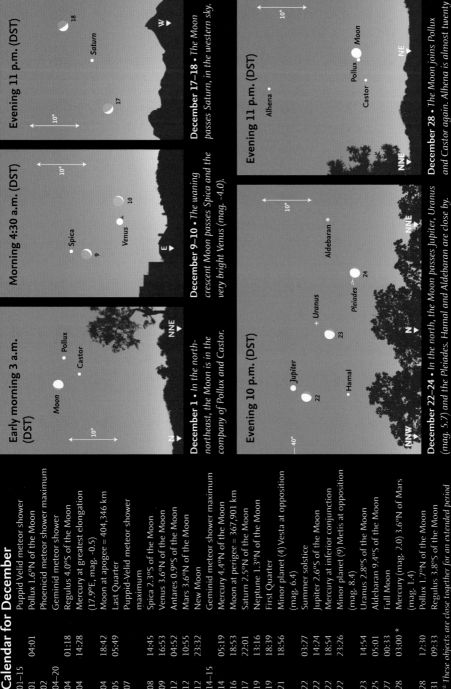

Calendar for December

01–15	Puppid Velid meteor shower
01	04:01 Pollux 1.6°N of the Moon
02	Phoenicid meteor shower maximum
04–20	Geminid meteor shower
04	01:18 Regulus 4.0°S of the Moon
04	14:28 Mercury at greatest elongation (17.9°E, mag. –0.5)
04	18:42 Moon at apogee = 404,346 km
05	05:49 Last Quarter
07	Puppid-Velid meteor shower maximum
08	14:45 Spica 2.3°S of the Moon
09	16:53 Venus 3.6°N of the Moon
12	04:52 Antares 0.9°S of the Moon
12	10:55 Mars 3.6°N of the Moon
12	23:32 New Moon
14–15	Geminid meteor shower maximum
14	05:19 Mercury 4.4°N of the Moon
16	18:53 Moon at perigee = 367,901 km
17	22:01 Saturn 2.5°N of the Moon
19	13:16 Neptune 1.3°N of the Moon
19	18:39 First Quarter
21	18:56 Minor planet (4) Vesta at opposition (mag. 6.4)
22	03:27 Summer solstice
22	14:24 Jupiter 2.6°S of the Moon
22	18:54 Mercury at inferior conjunction
22	23:26 Minor planet (9) Metis at opposition (mag. 8.4)
23	14:54 Uranus 2.8°S of the Moon
25	05:01 Aldebaran 9.4°S of the Moon
27	00:33 Full Moon
28	03:00 * Mercury (mag. 2.0) 3.6°N of Mars (mag. 1.4)
28	12:30 Pollux 1.7°N of the Moon
31	09:33 Regulus 3.8°S of the Moon

These objects are close together for an extended period around this time.

Evening 11 p.m. (DST)

December 17–18 • *The Moon passes Saturn, in the western sky.*

Morning 4:30 a.m. (DST)

December 9–10 • *The waning crescent Moon passes Spica and the very bright Venus (mag. –4.0).*

Early morning 3 a.m. (DST)

December 1 • *In the north-northeast, the Moon is in the company of Pollux and Castor.*

Evening 11 p.m. (DST)

December 28 - *The Moon joins Pollux and Castor again. Alhena is almost twenty degrees higher.*

Evening 10 p.m. (DST)

December 22–24 • *In the north, the Moon passes Jupiter, Uranus (mag. 5.7), and the Pleiades. Hamal and Aldebaran are close by.*

Dark Sky Sites

International Dark-Sky Association Sites

The **International Dark-Sky Association** (IDA) recognizes various categories of sites that offer areas where the sky is dark at night, free from light pollution and particularly suitable for astronomical observing. Although a number of sites are under consideration, the majority of confirmed sites are in North America. There more than 90 in the United States and Canada. There are 20 sites in Great Britain and Ireland, seven in Europe, and one each in Israel, Japan and South Korea. There are just 13 sites in the southern hemisphere.

Details of IDA are at: https://www.darksky.org/. Information on the various categories and individual sites are at:
https://www.darksky.org/our-work/conservation/idsp/.

Many of these sites have major observatories or other facilities available for public observing (often at specific dates or times).

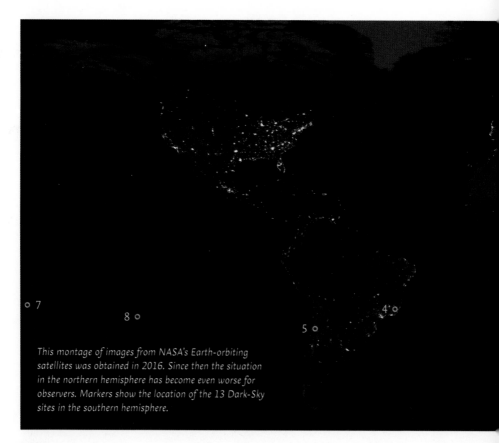

This montage of images from NASA's Earth-orbiting satellites was obtained in 2016. Since then the situation in the northern hemisphere has become even worse for observers. Markers show the location of the 13 Dark-Sky sites in the southern hemisphere.

1   *!Ae!Hai Kalahari Heritage Park* (South Africa)
2   *Aoraki Mackenzie* (New Zealand)
3   *Aotea / Great Barrier Island* (New Zealand)
4   *Desengano State Park* (Brazil)
5   *Gabriela Mistral* (Chile)
6   *NamibRand Nature Reserve* (Namibia)
7   *Niue* (New Zealand)
8   *Pitcairn Islands* (UK)
9   *River Murray* (Australia)
10 *Stewart Island / Rakiura* (New Zealand)
11 *The Jump-Up* (Australia)
12 *Wai-Iti* (New Zealand)
13 *Warrumbungle National Park* (Australia)

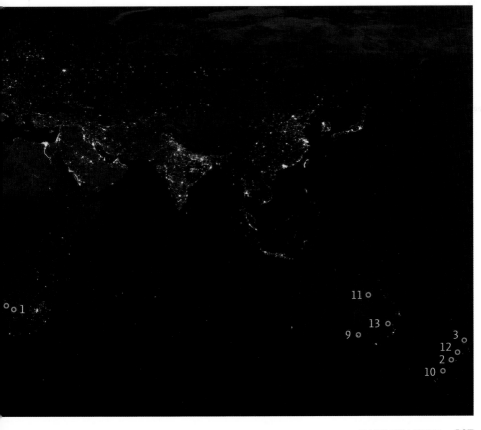

Glossary and Tables

aphelion	The point on an orbit that is farthest from the Sun.
apogee	The point on its orbit at which the Moon is farthest from the Earth.
appulse	The apparently close approach of two celestial objects; two planets, or a planet and star.
astronomical unit	(AU) The mean distance of the Earth from the Sun, 149,597,870 km.
celestial equator	The great circle on the celestial sphere that is in the same plane as the Earth's equator.
celestial sphere	The apparent sphere surrounding the Earth on which all celestial bodies (stars, planets, etc.) seem to be located.
conjunction	The point in time when two celestial objects have the same celestial longitude. In the case of the Sun and a planet, superior conjunction occurs when the planet lies on the far side of the Sun (as seen from Earth). For Mercury and Venus, inferior conjuction occurs when they pass between the Sun and the Earth.
direct motion	Motion from west to east on the sky.
ecliptic	The apparent path of the Sun across the sky throughout the year. Also: the plane of the Earth's orbit in space.
elongation	The point at which an inferior planet has the greatest angular distance from the Sun, as seen from Earth.
equinox	The two points during the year when night and day have equal duration. Also: the points on the sky at which the ecliptic intersects the celestial equator. The vernal (spring) equinox is of particular importance in astronomy.
gibbous	The stage in the sequence of phases at which the illumination of a body lies between half and full. In the case of the Moon, the term is applied to phases between First Quarter and Full, and between Full and Last Quarter.
inferior planet	Either of the planets Mercury or Venus, which have orbits inside that of the Earth.
magnitude	The brightness of a star, planet or other celestial body. It is a logarithmic scale, where larger numbers indicate fainter brightness. A difference of 5 in magnitude indicates a difference of 100 in actual brightness, thus a first-magnitude star is 100 times as bright as one of sixth magnitude.
meridian	The great circle passing through the North and South Poles of a body and the observer's position; or the corresponding great circle on the celestial sphere that passes through the North and South Celestial Poles and also through the observer's zenith.
nadir	The point on the celestial sphere directly beneath the observer's feet, opposite the zenith.
occultation	The disappearance of one celestial body behind another, such as when stars or planets are hidden behind the Moon.
opposition	The point on a superior planet's orbit at which it is directly opposite the Sun in the sky.
perigee	The point on its orbit at which the Moon is closest to the Earth.
perihelion	The point on an orbit that is closest to the Sun.
retrograde motion	Motion from east to west on the sky.
superior planet	A planet that has an orbit outside that of the Earth.
vernal equinox	The point at which the Sun, in its apparent motion along the ecliptic, crosses the celestial equator from south to north. Also known as the First Point of Aries.
zenith	The point directly above the observer's head.
zodiac	A band, streching 8° on either side of the ecliptic, within which the Moon and planets appear to move. It consists of twelve equal areas, originally named after the constellation that once lay within it.

The Constellations

There are 88 constellations covering the whole of the celestial sphere, but 4 of these in the northern hemisphere (Camelopardalis, Cassiopeia, Cepheus and Ursa Minor) can never be seen (even in part) from a latitude of 35°S, so are omitted from this table. The names themselves are expressed in Latin, and the names of stars are frequently given by Greek letters (see next page) followed by the genitive of the constellation name. The genitives and English names of the various constellations are included.

Name	Genitive	Abbr.	English name
Andromeda	Andromedae	And	Andromeda
Antlia	Antliae	Ant	Air Pump
Apus	Apodis	Aps	Bird of Paradise
Aquarius	Aquarii	Aqr	Water Bearer
Aquila	Aquilae	Aql	Eagle
Ara	Arae	Ara	Altar
Aries	Arietis	Ari	Ram
Auriga	Aurigae	Aur	Charioteer
Boötes	Boötis	Boo	Herdsman
Caelum	Caeli	Cae	Burin
Cancer	Cancri	Cnc	Crab
Canes Venatici	Canum Venaticorum	CVn	Hunting Dogs
Canis Major	Canis Majoris	CMa	Big Dog
Canis Minor	Canis Minoris	CMi	Little Dog
Capricornus	Capricorni	Cap	Sea Goat
Carina	Carinae	Car	Keel
Centaurus	Centauri	Cen	Centaur
Cetus	Ceti	Cet	Whale
Chamaeleon	Chamaeleontis	Cha	Chameleon
Circinus	Circini	Cir	Compasses
Columba	Columbae	Col	Dove
Coma Berenices	Comae Berenices	Com	Berenice's Hair
Corona Australis	Coronae Australis	CrA	Southern Crown
Corona Borealis	Coronae Borealis	CrB	Northern Crown
Corvus	Corvi	Crv	Crow
Crater	Crateris	Crt	Cup
Crux	Crucis	Cru	Southern Cross
Cygnus	Cygni	Cyg	Swan
Delphinus	Delphini	Del	Dolphin
Dorado	Doradus	Dor	Dorado
Draco	Draconis	Dra	Dragon
Equuleus	Equulei	Equ	Little Horse
Eridanus	Eridani	Eri	River Eridanus
Fornax	Fornacis	For	Furnace
Gemini	Geminorum	Gem	Twins
Grus	Gruis	Gru	Crane
Hercules	Herculis	Her	Hercules
Horologium	Horologii	Hor	Clock
Hydra	Hydrae	Hya	Water Snake
Hydrus	Hydri	Hyi	Lesser Water Snake
Indus	Indi	Ind	Indian
Lacerta	Lacertae	Lac	Lizard

Name	Genitive	Abbr.	English name
Leo	Leonis	Leo	Lion
Leo Minor	Leonis Minoris	LMi	Little Lion
Lepus	Leporis	Lep	Hare
Libra	Librae	Lib	Scales
Lupus	Lupi	Lup	Wolf
Lynx	Lyncis	Lyn	Lynx
Lyra	Lyrae	Lyr	Lyre
Mensa	Mensae	Men	Table Mountain
Microscopium	Microscopii	Mic	Microscope
Monoceros	Monocerotis	Mon	Unicorn
Musca	Muscae	Mus	Fly
Norma	Normae	Nor	Set Square
Octans	Octantis	Oct	Octant
Ophiuchus	Ophiuchi	Oph	Serpent Bearer
Orion	Orionis	Ori	Orion
Pavo	Pavonis	Pav	Peacock
Pegasus	Pegasi	Peg	Pegasus
Perseus	Persei	Per	Perseus
Phoenix	Phoenicis	Phe	Phoenix
Pictor	Pictoris	Pic	Painter's Easel
Pisces	Piscium	Psc	Fishes
Piscis Austrinus	Piscis Austrini	PsA	Southern Fish
Puppis	Puppis	Pup	Stern
Pyxis	Pyxidis	Pyx	Compass
Reticulum	Reticuli	Ret	Net
Sagitta	Sagittae	Sge	Arrow
Sagittarius	Sagittarii	Sgr	Archer
Scorpius	Scorpii	Sco	Scorpion
Sculptor	Sulptoris	Scu	Sculptor
Scutum	Scuti	Sct	Shield
Serpens	Serpentis	Ser	Serpent
Sextans	Sextantis	Sex	Sextant
Taurus	Tauri	Tau	Bull
Telescopium	Telescopii	Tel	Telescope
Triangulum	Trianguli	Tri	Triangle
Triangulum Australe	Trianguli Australis	TrA	Southern Triangle
Tucana	Tucanae	Tuc	Toucan
Ursa Major	Ursae Majoris	UMa	Great Bear
Vela	Velorum	Vel	Sails
Virgo	Virginis	Vir	Virgin
Volans	Volantis	Vol	Flying Fish
Vulpecula	Vulpeculae	Vul	Fox

The Greek Alphabet

α	Alpha	ε	Epsilon	ι	Iota	ν	Nu	ρ	Rho	φ (φ)	Phi
β	Beta	ζ	Zeta	κ	Kappa	ξ	Xi	σ (ς)	Sigma	χ	Chi
γ	Gamma	η	Eta	λ	Lambda	ο	Omicron	τ	Tau	ψ	Psi
δ	Delta	θ (ϑ)	Theta	μ	Mu	π	Pi	υ	Upsilon	ω	Omega

Some common asterisms

Belt of Orion	δ, ε and ζ Orionis
Cat's Eyes	λ and υ Scorpii
Circlet	γ, θ, ι, λ and κ Piscium
False Cross	ε and ι Carinae and δ and κ Velorum
Fish Hook	α, β, δ and π Scorpii
Head of Cetus	α, γ, ξ², μ and λ Ceti
Head of Hydra	δ, ε, ζ, η, ρ and σ Hydrae
Job's Coffin	α, β, γ and δ Delphini
Keystone	ε, ζ, η and π Herculis
Kids	ζ and η Aurigae
Milk Dipper	ζ, γ, σ, φ and λ Sagittarii
Pot	= Saucepan
Saucepan	ι, θ, ζ, ε, δ and η Orionis
Sickle	α, η, γ, ζ, μ and ε Leonis
Southern Pointers	α and β Centauri
Square of Pegasus	α, β and γ Pegasi with α Andromedae
Sword of Orion	θ and ι Orionis
Teapot	γ, ε, δ, λ, φ, σ, τ and ζ Sagittarii
Water Jar	γ, η, κ and ζ Aquarii
Y of Aquarius	= Water Jar

Acknowledgements

Denis Buczynski, Portmahomack, Ross-shire – p.32 (Fireball)

Dave Chapman, Dartmouth, Nova Scotia, Canada, p.77 (LMC)

European Southern Observatory, pp.53, 89, 101 (Magellanic Clouds)

Akira Fuji – p.21 (Lunar eclipse)

Bernhard Hubl – pp. 35, 43, 47, 49, 59, 65, 73, 83, 85, 91, 95, 103 (Constellation photographs)

Nick James, p.29 (Comet NEOWISE)

Arthur Page, Queensland, Australia, 41, 71 (Constellation photographs)

Damian Peach, Hamble, UK, p.23 (Mars photo)

peresanz/Shutterstock – p.37 (Orion)

Further Information

Books

Bone, Neil (1993), *Observer's Handbook: Meteors*, George Philip, London & Sky Publ. Corp., Cambridge, Mass.

Cook, J., ed. (1999), *The Hatfield Photographic Lunar Atlas*, Springer-Verlag, New York

Dunlop, Storm (1999), *Wild Guide to the Night Sky*, HarperCollins, London

Dunlop, Storm (2012), *Practical Astronomy*, 3rd edn, Philip's, London

Dunlop, Storm, Rükl, Antonin & Tirion, Wil (2005), *Collins Atlas of the Night Sky*, HarperCollins, London

Ellyard, David & Tirion, Wil (2008), *Southern Sky Guide*, 3rd edn, Cambridge University Press, Cambridge

Heifetz, Milton & Tirion, Wil (2017), *A Walk through the Heavens*, 4th edn, Cambridge University Press, Cambridge

Heifetz, Milton & Tirion, Wil (2012), *A Walk through the Southern Sky*, 3rd edn, Cambridge University Press, Cambridge

O'Meara, Stephen J. (2008), *Observing the Night Sky with Binoculars*, Cambridge University Press, Cambridge

Ridpath, Ian (2018), *Star Tales*, 2nd edn, Lutterworth Press, Cambridge, UK

Ridpath, Ian, ed. (2003), *Oxford Dictionary of Astronomy*, 2nd edn, Oxford University Press, Oxford

Ridpath, Ian, ed. (2004), *Norton's Star Atlas*, 20th edn, Pi Press, New York

Ridpath, Ian & Tirion, Wil (2004), *Collins Gem – Stars*, HarperCollins, London

Ridpath, Ian & Tirion, Wil (2017), *Collins Pocket Guide Stars and Planets*, 5th edn, HarperCollins, London

Ridpath, Ian & Tirion, Wil (2019), *The Monthly Sky Guide*, 10th edn, Dover Publications, New York

Rükl, Antonín (1990), *Hamlyn Atlas of the Moon*, Hamlyn, London & Astro Media Inc., Milwaukee

Rükl, Antonín (2004), *Atlas of the Moon*, Sky Publishing Corp., Cambridge, Mass.

Scagell, Robin (2000), *Philip's Stargazing with a Telescope*, George Philip, London

Sky & Telescope (2017), *Astronomy 2018*, Australian Sky & Telescope, Quasar Publishing, Georges Hall, NSW

Stimac, Valerie (2019) *Dark Skies: A Practical Guide to Astrotourism*, Lonely Planet, Franklin, TN

Tirion, Wil (2011), *Cambridge Star Atlas*, 4th edn, Cambridge University Press, Cambridge

Tirion, Wil & Sinnott, Roger (1999), *Sky Atlas 2000.0*, 2nd edn, Sky Publishing Corp., Cambridge, Mass. & Cambridge University Press, Cambridge

Journals

Astronomy, Astro Media Corp., 21027 Crossroads Circle, P.O. Box 1612, Waukesha, WI 53187-1612 USA. http://www.astronomy.com

Astronomy Now, Pole Star Publications, PO Box 175, Tonbridge, Kent TN10 4QX UK. http://www.astronomynow.com

Sky at Night Magazine, BBC publications, London. http://skyatnightmagazine.com

Sky & Telescope, Sky Publishing Corp., Cambridge, MA 02138-1200 USA. http://www.skyandtelescope.com/

Societies

British Astronomical Association, Burlington House, Piccadilly, London W1J 0DU. http://www.britastro.org/

The principal British organization for amateur astronomers (with some professional members), particularly for those interested in carrying out observational programmes. Its membership is, however, worldwide. It publishes fully refereed, scientific papers and other material in its well-regarded journal.

Federation of Astronomical Societies, Secretary: Ken Sheldon, Whitehaven, Maytree Road, Lower Moor, Pershore, Worcs. WR10 2NY. http://www.fedastro.org.uk/fas/

An organization that is able to provide contact information for local astronomical societies in the United Kingdom.

Royal Astronomical Society, Burlington House, Piccadilly, London W1J 0BQ. http://www.ras.org.uk/
The premier astronomical society, with membership primarily drawn from professionals and experienced amateurs. It has an exceptional library and is a designated centre for the retention of certain classes of astronomical data. Its publications are the standard medium for dissemination of astronomical research.

Society for Popular Astronomy, 36 Fairway, Keyworth, Nottingham NG12 5DU.
http://www.popastro.com/
A society for astronomical beginners of all ages, which concentrates on increasing members' understanding and enjoyment, but which does have some observational programmes. Its journal is entitled *Popular Astronomy*.

Software

Planetary, Stellar and Lunar Visibility (planetary and eclipse freeware): Alcyone Software, Germany.
http://www.alcyone.de
Redshift, Redshift-Live. http://www.redshift-live.com/en/
Starry Night & Starry Night Pro, Sienna Software Inc., Toronto, Canada. http://www.starrynight.com

Internet sources

There are numerous sites with information about all aspects of astronomy, and all of those have numerous links. Although many amateur sites are excellent, treat any statements and data with caution. The sites listed below offer accurate information. Please note that the URLs may change. If so, use a good search engine, such as Google, to locate the information source.

Information

Astronomical data (inc. eclipses) HM Nautical Almanac Office: http://astro.ukho.gov.uk
Auroral information Michigan Tech: http://www.geo.mtu.edu/weather/aurora/
Comets JPL Solar System Dynamics: http://ssd.jpl.nasa.gov/
American Meteor Society: http://amsmeteors.org/
Deep-sky objects Saguaro Astronomy Club Database: http://www.virtualcolony.com/sac/
Eclipses NASA Eclipse Page: http://eclipse.gsfc.nasa.gov/eclipse.html
Ice in Space (Southern Hemisphere Online Astronomy Forum): http://www.iceinspace.com.au/index.php?home
Moon (inc. Atlas) Inconstant Moon: http://www.inconstantmoon.com/
Planets Planetary Fact Sheets: http://nssdc.gsfc.nasa.gov/planetary/planetfact.html
Satellites (inc. International Space Station)
Heavens Above: http://www.heavens-above.com/
Visual Satellite Observer: http://www.satobs.org/
Star Chart http://www.skyandtelescope.com/observing/interactive-sky-watching-tools/interactive-sky-chart/
What's Visible
Skyhound: http://www.skyhound.com/sh/skyhound.html
Skyview Cafe: http://www.skyviewcafe.com

Institutes and Organizations

European Space Agency: http://www.esa.int/
International Dark-Sky Association: http://www.darksky.org/
RASC Dark Sky: https://www.rasc.ca/dark-sky-sites/
Jet Propulsion Laboratory: http://www.jpl.nasa.gov/
Lunar and Planetary Institute: http://www.lpi.usra.edu/
National Aeronautics and Space Administration: http://www.hq.nasa.gov/
Solar Data Analysis Center: http://umbra.gsfc.nasa.gov/
Space Telescope Science Institute: http://www.stsci.edu/